AF553908

Water Pollution

Water Pollution

S. K. Agarwal

A P H PUBLISHING CORPORATION
4435-36/7, ANSARI ROAD, DARYA GANJ
NEW DELHI-110 002

Published by
S.B. Nangia
APH Publishing Corporation
4435-36/7, Ansari Road, Daryaganj
New Delhi 110002
Ph.: 23274050
E-mail : aphbooks@gmail.com

2026

Rs. 2995/-

Printed at
Balaji Offset
Navin Shahdara, Delhi 110032

SERIES PREFACE

This series of books on pollution management, is the outcome of difficulties encountered by the author while imparting the knowledge of the essential elements of pollution management know-how to the students. There is so much diversification of technical details that even teachers find difficulty in preparing regular lectures in a proper continuum. An effort has been made to provide all aspects of pollution management that are included in different syllabi of the Indian Universities.

Though the author do not claim any originality in explaining the various topics yet it has been tried to be genuine in arrangement, style and method of explanation of the subject matter. These aspects make the series most suitable for understanding the basic concepts of pollution management. However, the teachers and students would be able to judge how far the author have been successful in his efforts.

This series of books presents an authoritative explanation and discussion of a wide range of problems related to the environment, at a level suitable for practitioners and students in science, engineering, medicine, administration and planning. For the increasing number of teachers and students involved in degree and diploma courses in environmental science. The series should be particularly useful.

PREFACE

Water is one of the most important resource which human being has exploited than any other resource for the sustenance of their life. Most of the water on this planet is stored in oceans and ice-caps which is difficult to be recovered for our diverse needs. Most of our demands for water are fulfilled by rainwater which gets deposited in surface and groundwater sources. The quantity of this utilizable water is very much limited on the earth. Though water is continuously purified by evaporation and precipitation, pollution of water has emerged as one of the most significant environmental problem. Hot only there is an increasing concern for rapidly deteriorating water quality, but the quantity of unutilisable water is also diminishing. The causes of such a situation are many, but gross pollution of water has its origin in urbanisation, industrialisation and increasing human population.

Our fresh water lakes, rivers and oceans have self-correcting mechanisms, which check the excess of one thing over the other. Human interference was also taken care of by the nature itself, until modern technology created a situation where excessive exploitation has not only polluted our water resources, but posed a serious threat to the very existence of life. Water pollution management therefore, is not a mere slogan, but a dire necessity and unless a serious attention is paid to it, we may reach a point of no return.

The frontiers of water pollution are not east or west, north or south, but wherever one looks for them - and it is gratifying to note that many different persons, inlcuding scientists, sociologists, politicians, economists, geographers, geologists and hydrologists are now attempting to seriously look at these frontiers from diverse angles. This book is based on recent views, ideas and contributions of some of the world's leading ecologists, with special reference to comprehensive information on water pollution, regarding their source, effects and control. Separate chapters have, therefore, been provided for Introduction, water pollution, thermal pollution, inorganic pollution, organic pollution, oil pollution, sewage pollution, aquatic pathogens and nuisance organisms, eutrophication, water quality standards. Biotechnology for water pollution control, water quality management, references and subject index. These chapters form an important aspect of water pollution management. For abatement of water pollution, the treatment of wastewater is considered to be the best strategy. Some of the common methods used for wastewater treatment, including sewage treatment and drinking water purification, have also been discussed in this book.

I am thankful to my colleagues and friends for their suggestions, discussions and criticisms. I thank Prof.L.N.Vyas, Udaipur who provided useful suggestions, and whose works have been freely quoted here in preparation of this book. I am particularly grateful to Prof. R.R. Das, Gwalior and Prof. P.S. Dubey, Ujjain for providing very useful information on this subject. I am thankful to Prof. G.J. Bakus, University of Southern California, Los Angeles for sending me his observations on knowledge and excitement on marine deterioration. I also express my sincere appreciation to Prof. Jun Ui, Tokyo University, Japan for extending literature on mercury pollution. The active support and forbearance shown by my wife for a long period of time was, perhaps, the major factor is showing this book the light of the day.

Dr. S.K. Agarwal

CONTENTS

Chapter-1

INTRODUCTION

Water is a "cradle of life" on which all organisms play. Water makes upto 60 - 95 percent of weight of any functioning organism and an equal or even greater proportion of the weight of any living cell. It is a *common* liquid, by far the most common. But common should not be mistaken with *ordinary;* water is not in the least an ordinary liquid. In fact, it is, extraordinary in comparison with other liquids. If it were not, it is most probable that life - at least, our form of life, could never have evolved.

Water is perhaps one of the most peculiar of our natural resources for life, next to air it is likely to become a critical scarce resource in the coming decades. While we consider it a renewable resource, in the sense that it is possible for us to control it, regulate its use and improve its quality and insure that we will have a continuing supply of it, it is a non-renewable resource in the sense that for the entire earth there is a fixed and relatively unchanged amount.

The amount of water which is present on the earth is vast. Most of the earth's surface is covered with water. The water which is present today was formed eons age during a time when our earth was changing from a molten mass to essentially the form in which we know it today. During the cooling period chemical changes occurred, gases were released and from these gases water was formed. This water has appeared at different times in different ways and has covered varying amounts of the earth. But, over all of these changes and times and places the amount of water has remained essentially the same.

About 70 percent of the world's surface area is covered with water. Yet only 0.00192 percent of the total stock of water is available for human use because 98 percent of water in oceans and seas and 1.998 percent locked up in arctic regions, glaciers, mountains and clouds is not available. About 85 percent of the rainfall's directly into the sea and never reaches the land. The remaining rainfall reaching the land fills the lakes, ponds, underground water resources and keeps the river flowing. The latter constitutes only 0.00008 percent of the total water. Thus, humanity is left with only one teaspoonful of sweet water for every 5 litres of total water (CPCB, 1978-79).

For much of the earth's early period water covered even more of this surface than it does at the present time. In more recent periods of the earth's history large quantities of water were in the form of flaciers which covered great areas of the earth's surface. Today we are living in a period which many geologists believe is between glacial epoches. The great polar ice caps and alpine glaciers seem to be melting and releasing more water into the oceans.

While these long term changes in the status of water on the earth's surface are interesting they have little significant effect on the day-to-day problems of water management. It is the more temporal variations in water abundance, location and quality that need to be understood if we are to manage out water resources properly.

Water is one of the most important resource which human being has exploited than any other resource for the sustenance of his life. Water is the substance which is present in all the three states of matter, that is, gaseous (water vapours), liquid, and solid (ice) within the ranges of temperature and pressure common to the earth. The distribution of water in its various forms on the earth (Table 1) show that approximately 97 percent of the total water exists in gigantic oceans. This water is of little importance in our daily requirement. Remaining 3 percent of water is distributed in the form of ice sheets; underground sorces; lakes, rivers, ponds (surface sources); atmosphere (water vapours), and biological water contained in the living organisms. The surface fresh water in the form of lakes and rivers is hardly 0.01 percent of the total water on the earth. Since our demand for water are mostly for fresh water, we have to depend mainly on the tiny fraction of the total water present on this planet. Further, the uneven distribution of water on the surface of the earth, makes it a scarce resource at several places.

Table 1. Distribution of water on the earth.

Component	km^3	Geograms(G)	%
Oceans	1300,000,000	13000	96.42
Ice sheets	24000.000	240	1.78
Underground	24000,000	240	1.78
Lakes	180.000	1.8	0.01
Atmosphere	13000	0.13	0.001
Rivers	2000	0.02	0.00015
Biological water (Water contained in living organisms)	1000	0.01	0.000075

$G = 1 \times 10^{20}$ g

Observations (Table 2) indicates that maximum quantity of water occurs in rocks, followed by the oceans, while minimum quantity of water occurs in the atmosphere as water vapours. Every year 4.46 G of water comes in the form of rainfall of which 3.47 G precipitates over the ocean's surface. About 1 G rainfall occurs over landmass of which 0.2 G runs away and 0.6 G evaporates again, and only a small quantity (0.2 G) is stored as underground water, 0.13 G water moves in the form of water vapours and clouds from ice caps present on South and North poles and on the top of hifh mountains. Only about 0.004 percent (~ 10 G) of the total water is all the time moving in the cycle as much of the earth's water is in cold storage. Glaciers and ice caps cover 11 percent of the world's land area; permanently frozen ground holds another 10 percent in its grip, while 30-50 percent of the land is covered with snow at any given time. Icebergs and pack ice occupy 25 percent of the oceans area. Therefore, all freshwater is locked up as ice, mostly in the Antarctica and Greenland.

Table 2. Distribution of water on the earth (Hutchinson, 1957).

Water in the rocks (lithosphere)	250000.00 G
Water in sedimentary rocks	2100.00 G
Water in the oceans	13800.00 G
Water in rivers and lakes	0.25 G
Water in atmosphere as vapours	0.13 G
Water in mountainous and polar ice caps	167.00 G
Circulating ground water	2.50 G
Total quantity of water on the earth	266069.88 G

Geogram (G) - 10^{20} g

Surface, subsurface and atmospheric water, in all its forms, is collectively called as hydrosphere. Surface water refers to oceans, lakes, rivers, glaciers and polar icecaps. Subsurface water consists of ground water including soil moisture. Atmospheric water may occur in gaseous state as water vapour, as liquid (rain drop, clouds, and fog) or as a solid (snow or ice).

OCEANS

The oceans on the earth cover some 362 million square kilometer area, and are the sink into which all water finally flows. The water flowing over the land picks up a number of elements and compounds from the soil and air, transports these substances in solution and deposits them in the oceans. Since there is no exit for these dissolved substances, they tend to accumulate in ocean water. The average total salt content of the ocean is 35 percent (35 gm per kg of water). The most concentrated substance in the oceans is Sodium chloride, which makes nearly 90 percent of the total dissolved salts. Ocean water contains about 57 elements in solution.

The oceans cannot be considered primary resources for the direct water supply for domestic, agricultural or industrial use owing to their high salt content. However, at some places, the ocean water is being used for these purposes after the process of desalination.

The oceans are very useful resource for obtaining other commodities like food (fish and other sea food), common salt and other substances. In certain areas, the electrical energy is also obtained from the oceans as tidal energy and hydrothermal energy.

GROUND WATER

The volume of ground water is much greater than that of all freshwater lakes and streams combined. Underground water plays an important role in the overall water balance of the earth. As a reservoir, it has an enormous capacity to store water in rainy season which can be utilized during dry periods. It helps in controlling the river flow.

Ground water is a primary source of freshwater in several urban and rural areas. The existing utilization of groundwater in India is about 45,000 million cubic meters. It is widely used as a source of water for domestic, irrigation and other farm uses.

Ground water are recharged by atmospheric precipitation. Problems may result if the amount withdrawn from ground water sources is greater than that is replanished by rainfall. The water table may go down substantially by over exploitation. Near the coastal areas, over exploitation will lead to the flow of marine water into freshwater aquifers.

LAKES AND RESERVOIRS

Lakes, ponds and reservoirs represent the lentic waters (stagnant waters). They serve for the variety of purposes like domestic water supplies, industrial processes, irrigation, navigation, fish and water fowl, flood protection, recreation and generation of hydroelectricity. In the rural areas or small communities ponds are important source of water for cloth-washing, bathing and for cattle.

In India, there are several large existing multipurpose reservoirs besides several large natural lakes and innumerable small tanks and ponds. Dam is one of the easiest device to store water. The discharge of the reservoir can be controlled according to the demand . The flow of the river in downstream can be augmented during dry periods by releasing extra water from the dam to ensure a continuous water supply. The construction of dam in certain upstream region can check the floods in the downstream region during the rainy season.

However, there are certain ecological problems inherent in the reservoir construction. The evaporation of water increases several times due to the enlargement of the surface area of water. The aquatic weeds infest the reservoir, impairing the water quality, and further reducing the total quantity of water by greater evapo-transpiration. These weeds should be continuously removed from the water in order to keep the reservoir in healthy condition. To check the surface evaporation, especially in the drier regions, surface film of certain chemicals are applied. These films must be capable of allowing the passage of oxygen and carbon dioxide, non-toxic and self-restoring after disturbance.

RIVERS

Rivers have been considered to be one of the most important water resource. India has a net of rivers (Table 3) with 14 major river basins, with 83 percent of total land area covered, 85 percent contribution of total surface flow, they have 80 percent of total population. The 14 major river basins are Indus, Ganga, Brahamputra, Sabarmati, Mahi, Narmada, Tapti, Subarnerekha, Brahmani, Mahanadi, Godavari, Krishna, Pennai and Cauvery. Out of the 14 major rivers, 4 are perennial having an annual average minimum discharge of $0.47 \times 10^6 m^3/km$, 6 have $0.26 \times 10^6 m^3/km$ and the remaining 4 have between 0.06 and $0.24 \times 10^6 m^3/km$. Except the 4 major rivers Brahamputra, Ganga, Indus and Godavari the rest carry very little water in summer months. Being rainfed, the flow corresponds to the rainfall. During the monsoon (3-4 months), rivers flow to their brim but in other periods, the flow dwindles and several rivers are practically dry during summer when the demand for water is highest and water pollution is acute. This results in the irony of inadequate water resources in a country with an adequate rainfall.

Table 3 . Classification of Indian rivers and their estimated discharge potential (Rao, 1975; Nag and Kathpali, 1975).

Category of rivers	No. of rivers	Annual discharge (10^{12} m^3)	% of total dis-charge	Each river with a catchment area (km^2)	% of total land area covered	% of total population living in the area
Major	14	1.406	85	20,000 & above	83	80
Medium	44	0.112	7	2,000 to 20.000	8	20
Minor & desert	55	0.127	8	Less than 2,000	9	
Flow from outside India	-	0.200	-	-		

The indiscriminate and largescale deforestation and overgrazing in the watershed areas of river basins have caused considerable siltation of dams and shrinking of river flow. This leads to the flooding of rivers at the times of excessive rains.

ATMOSPHERIC WATER

All water in the environment passes through the atmosphere at some time during the hydrologic cycle. The atmospheric water is important in determining the conditions in the environment that affect our life supporting system.

Though atmospheric water is not considered as a direct source of water supply, it is the ultimate source of all fresh water supplies on the earth that come in the form of precipitation. The quality of this water in considered to be purest.

Although the average precipitation over the country as a whole is about 1000 mm, this is very unevenly distributed in space and time. The west-coast and the Assam region are areas of heavy rainfall, receiving 2500 mm and above annually. The eastern part of the peninsula and the northern plains receive moderate rainfall of 1000-2500 mm annually. The Punjab plains and upper western part of the Deccan plateau receive low rainfall of 250-1000 mm.

While the Rajasthan desert and Ladakh plateau of Jammu and Kashmir are regions of very low precipitation of less than 250 mm. The bulk of the precipitation occurs in the south west monsoon period covering 4 to 5 months of June to October. A large part of the country experiences acute water shortage in the other months. It is only the south-eastern coast of peninsular India that receive the major share of the precipitation in November and December from the north-east monsoon (Murthy, 1975).

An important function of the atmospheric water is to keep the temperature moderate. Water in atmosphere makes it possible for the air to absorb and hold vast amount of heat. Because of the presence of water vapour, the temperature of the atmosphere remains within the hospitable range for most of the living organisms.

HYDROLOGICAL CYCLE

Water is not locked permanently in the various components of the earth but it has a unique feature among most of the resources in that it is constantly in a cycle. The illustration (Figure 1) describes the basic processes of that cycle, wherein water from the surface of the oceans and land evaporates, enters the atmosphere, precipitates, falls upon the land and runs into the ground or on the surface of the ground down hill through streams and underground aquifers to the oceans where once again it is evaporated and enters the atmosphere. This cycle is in continuous flow. Therefore, man's use of water essentially involves withdrawals from that portion of the cycle when the water is flowing on top of and through the upper layers of the earth's surface. Water in this condition is practically speaking, free of salts, and is in a form which is needed by all terrestrial forms of plant and animal life.

Solar energy evaporates the water from oceans, rivers and lakes into the atmosphere where it forms the clouds. The winds transports these clouds to various parts of the earth. The vapours in the clouds condense and precipitates in the form of either dew, rain, snow or hail on the earth. A large part of the precipitation takes place over the oceans themselves, while the remaining precipitates on the land masses. The water falling on the land masses is the potential supply which is determined by the routes

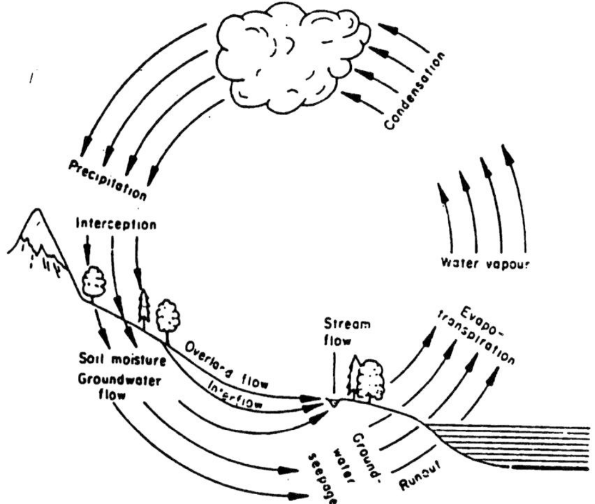

Fig.1 . The hydrologic cycle (after Ward, 1973).

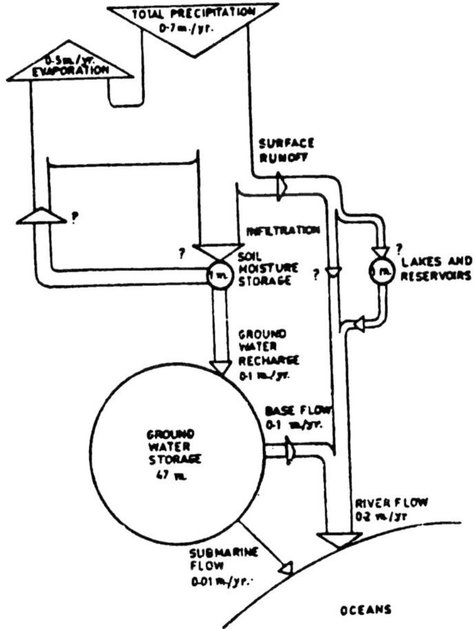

Fig.2. The hydrologic cycle (after Gehlar, 1972).

by which it is again retuned to the atmosphere. Of the water falling on the land surface, some is evaporated again and some is returned by surface run-off to drains, streams, rivers and lakes reaching finally to the oceans. The remaining water is infiltrated into the soil, and percolates deep into the ground water from where part of it may seep into streams, lakes or directly to the oceans. Some water on the land is absorbed by plants and consumed by the animals. This water, however, is released again into the atmosphere by respiration and evapo-transpiration. The global water cycle operates at a rapid rate with average 10 days residence time of water vapours into the atmosphere. The total global water is regarded to be present in a series of storage tanks interconnected by the transfer processes of evaporation, moisture transport, condensation, precipitation and run-off.

Figure 2 represents the hydrological cycle in terms of average global conditions on land, with the rate of flow and amounts of storage indicated by the width of flow, lines and areas of circles respectively. The implied average time for a drop of water to travel through the ground water system, from rainfall to the seas, is somewhat over 400 years. This is because ground water spends long time moving through very small passages in the soil.

As regards to the freshwater resources on the earth, precipitation on the land masses is critical since, out of the total precipitation, about 75 percent falls directly on the oceans, and only 25 percent comes to the land surface. The distribution of this precipitation on land is highly uneven, therefore, the pattern of natural flow may be an important factor from the resource point of view. In fact, the water management practices are based on the manipulation of hydrological cycle on a local scale.

HUMAN INTERFERENCE WITH HYDROLOGICAL CYCLE

Human beings influences the hydrological cycle in various ways. The impact on the global scale is little because of the inability of human beings to control the energy distribution on the earth and global climate which are the main forces in governing the hydrological cycle. However, human beings have scored a good success in manipulating the cycle by controlling the processes like run-off, evaporation, precipitation and infiltration of water on a local scale.

The deforestation is sometimes carried out in the catchment or watershed areas of the basin to increase the water yield by augmenting the run-off. However, this, unfortunately decreases the infiltration of water into the soil, as there is lack of the obstacles and organic matter which facilitate capturing of the surface moisture. The direct heavy rainfall on the bare surface may lead to the pulverization of surface materials which block the soil pores and reduce infiltration.

The agricultural field where the soil is not covered throughout the year also promote run-off. Deforestation increases the soil water because the reduction in the consumption of water by the plants. In some areas, where heavy precipitation is associated with steep slopes, floods may occur with the decrease of time lag between precipitation and run-off. The percolation and run-off depends mainly on the soil types. In sandy or permeable soils, though percolation is more, the upper loose profile of soil is lost with the run-off along with the nutrients. This process is called soil erosion. In the clays or less permeable soils, soil saturation may be caused with development of anaerobic conditions, particularly in the cold and humid regions.

In the areas where precipitation is low with marked seasonality and variability with higher evaporation rates, the manipulation of hydrological cycle may result in the development of arid conditions, The run-off is increased by the removal of vegetation at the expense of infiltration, the water left in the soil evaporates, and the underground water is tapped beyond the quantity recharged. All this leads to the gradual scarcity of water in the area. As the water table goes down, the wells are further deepened. Ground water becomes non-renewable resource as the withdrawal increases beyond the recharge. Rivers and streams get dried up fast because of the lowering of water table. Continued cropping and grazing increase the risk of wind erosion, and symptoms of drought become pronounced. Eventually, the productive areas turn into deserts.

Human intervention in changing the course of rivers and streams and divertinf them to the areas of need, may cause enormous ecological problems. The creation of dams for storage of flowing waters is also of ecological concern. These interventions increases the natural evaporation of water by increasing the surface area of water. Dams can induce earthquakes as has been reported in several parts of the country and the world, The downstream quality of water is also affected and there may be some ecological consequences where the river meets sea. Our intervention, to augment the flow of rivers in lean seasons by increased melting of ice covers on the mountains, may alter the ecology and microclimate of the area.

Artificial rainfall which is carried out by cloud seeding, is another aspect of interference with the hydrological cycle. Our success in this field is limited to only a small scale. In time to come, when effective methods will be developed and applied on a large scale, the result may be disastrous floods in some areas and/or serious drought in another areas.

WATER BALANCE

The existence of hydrological cycle is responsible for ecosystem water balance. Figure 3 shows the path of water through the system when precipitation exceeds transpiration and evaporation. Rain may pass the canopy as thro-fall or be intercepted by foliage from which it may either evaporate directly or be channelled into stem-flow or leaf-drip and increase the input to the soil surface. Infiltration replanishes the soil storage reservoirs or leads to discharge. If the infiltration capacity is exceeded, lateral run-off may occur, reducing the input to some parts of the soil, and increasing it to others. Lateral movement below the water table may also be a source of income or loss. Water uptake by plants and its loss, by transpiration, represents the main return pathway to the atmosphere, reinforced by some evaporation from the soil surface. Between periods of rain and soil storage capacity buffers these changes and may be replanished by apillary rise from the water table. Also shown is the intervention of the dominant directions of water movement when evaporation exceeds precipitation. Much less water enters the soil and capillary rise may be all important both in supplying the plant with water and in determining the pedogenic processes. Surface enrichment with solutes may occur compared with the strong leaching processes.

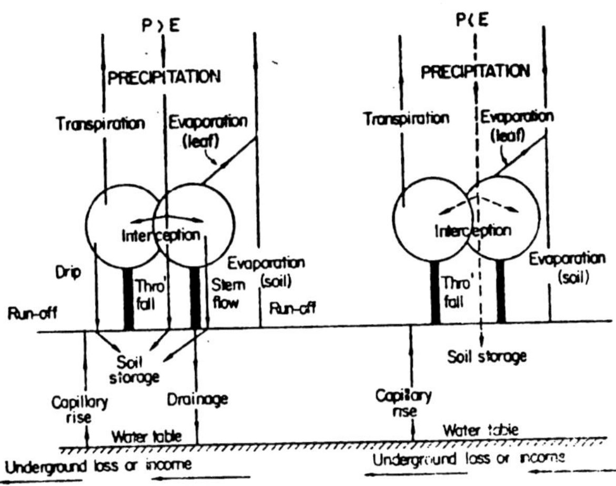

Figure 3 . Ecosystem water balance (Ethrington, 1975).

The hydrological cycle shows that the input of water in any component of the biosphere should be equal to output. On the land surface the water balance equation is as follows:

$$r = E + f_w + G$$

where r = rainfall
E = evaporation
f_w = surface run-off
G = flow of moisture from the earths surface to lower layers

The amount of moisture as G can be divided as the percolation in the soil (f_p) and changes in the moisture content in the upper layers of the lithosphere (soil moisture and aquifers, b). If we combine surface run-off (f_w) and percolation in the soil (f_p) as *f*, the water balance equation can be written as:

$$r = E + f + b$$

For an average yearly period b is relatively small and can be ignored, hence the water balance equation will become:

$$r = E + f$$

The yearly water balance of various continents and the whole land have been shown in Table 4.

Table 4. Water balance on land masses (Budyko, 1980).

Continent	Precipitation cm/year	Evaporation cm/year	Run-off cm/year
Europe	77	49	28
Asia	63	37	26
Africa	72	58	14
North America	80	47	33
South America	160	94	66
Australia	45	41	4
All land average	80	48.5	31.5

In calculating the water balance on the whole earth, there is no significance of the distribution of water, and the precipitation should essentially be equal to evaporation on the yearly basis:

$$r = E$$

The water balance on the earth has been schematically shown in Figure 4.

The total evaporation from the land (80 Tm/year) and oceans (420 Tm/year) is about 500 Tm/year. The total precipitation on the earth is also 500 Tm/year out of which 380 Tm/year falls over the oceans and rest 120 Tm/year on the land masses. This shows that there is net gain of 40 Tm/year of water on the land masses which comes from the oceans. This extra 40 Tm/year of water is redistributed and flows to oceans through run-off and percolation.

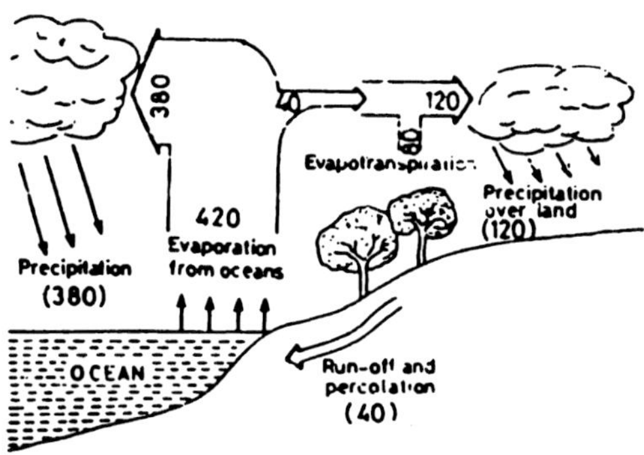

Fig 4. Annual water budget on the earth. Data based on Dix (1981) (Values are in Tm/Year. Tm = Tera metre cube = 10^{12} m^3).

STRUCTURE OF WATER

Each molecule of water (H_2O) consists of two hydrogen atoms and one oxygen atom. Each hydrogen atom is united to oxygen atom by sharing a pair of electrons. As the oxygen atom has eight electrons and hydrogen atom has one electron, in a water molecule, three nuclei are sorrounded by ten electrons.

The two electrons of Is orbital of oxygen are confined to the vicinity of oxygen nucleus, while the remaining eight form the four other orbitals (Figure 5). Two of the four orbitals are directed along O-H bond, where as the other two with unshared electron pair in each are projected above and below the H-O-H molecular plane. These two unshared pair of electrons provide strong negative regions on the oxygen atom in the water molecule. As the oxygen nucleus has eight protons in its nucleus, therefore, it has greater tendency to attract the shared electrons which leaves an intense positive electric field on the hydrogen atoms capable of attracting the unshared pairs of the oxygen atom. The bond thus developed by the attraction between positively charged hydrogen atom of one water molecule and negatively charged oxygen atom of another water molecule is termed as hydrogen bond (Figure 6).

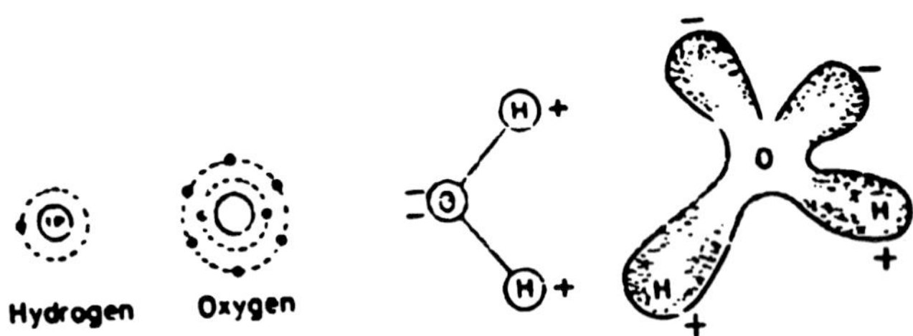

Fig. 5. Atomic structures of hydrogen and oxygen, and molecular structure of H_2O showing development of polarity.

Fig. 6. Formation of hydrogen bonds between water molecules.

These hydrogen bonds are responsible for joining several water molecules together and forming a three-dimensional structure. Most of the anomalous properties of water can be explained on the basis of the presence of hydrogen bonds. In the liquid water, there are average 2-3 hydrogen bonds per molecule of water, but in the ice four bonds per molecule in ice gives the ice a three-dimentional figure with each water molecule is surrounded by four water molecules in a tetra tetrahedral fashion to form a tridymite arrangement with the oxygen atoms forming layers of "puckered" hexagonal rings (Figure 7). The hexagonal arrangement of ice gives an open structure, in which the water molecules are less compacted, thereby explaining the low density of ice as compared to liquid water.

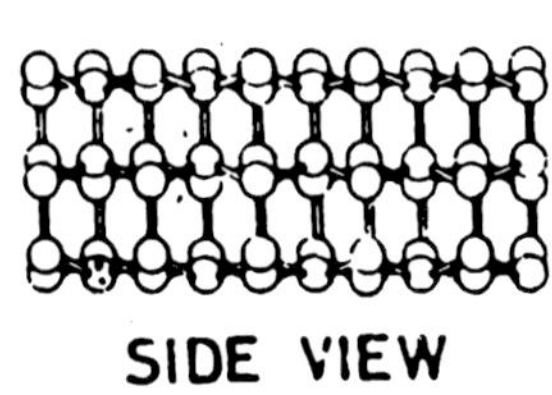

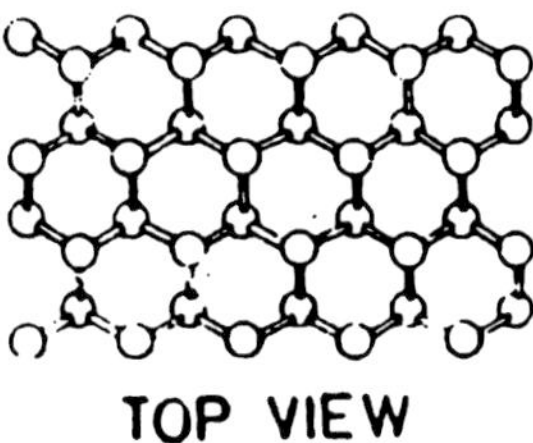

Fig. 7. Structure of ice showing arrangement of water molecules.

PHYSICAL PROPERTIES OF WATER

Water is an extremely inert body in relation to most other chemical substances. It contains following unique physical properties which have ecological relevance :

(a) Specific heat

Water is capable of storing tremendous quantities of heat energy with a relatively small rise in temperature. Water has a specific heat of 1.0 calorie which is highest among all substances except lithiurn, ammonia and liquid hydrogen. Thus, water is said to have a high specific heat which can be defined as the *number of calorie necessary* to *raise the one gram of water* to *one degree centigrade.* Large bodies of water do not fluctuate much in temperature as compared to large bodies of land, because so much heat is required to raise the water temperature. This property of water enables shoreline areas to have relatively more constant temperatures than the inland areas, because such great quantities of heat are absorbed before the temperature of natural waters such as ponds, lakes and seas can be raised to one degree centigrade, they warm up slowly in the spring and cool off just as slowly in the fall. This tends to maintain a relatively constant temperature of aquatic habitats and prevents wide seasonal fluctuations in the temperature of ponds, lakes and seas and also moderates the temperatures of local and worldwide environments.

(b) Latent heat

Water possesses the highest heat of fusion and heat of evaporation, collectively called latent heat. Thus, large quantities of heat energy (79.71 cal per gram) must be removed before water can change from a liquid to a solid state, and conversely, it must absorb considerable heat before ice can be converted to the liquid state. It takes approximately 80 calories of heat to convert one gram of ice to a liquid state when both are at 0°C. This is equivalent to the amount of heat needed to raise the same quantity of water from 0°C to 80°C.

Evaporation occurs at the interface between air and water at all ranges of temperature. It also requires considerable amount of heat, 539.55 cal of heat are needed to overcome the attraction between molecules and convert one gram of water at 100°C into vapour.

The two types of latent heats are because of the presence of hydrogen bonds. Excess heat is required to break these bonds at the time of changing the state of water, that is, from liquid to vapour and from ice to liquid. These properties help in moderating the surface temperature of large water bodies. Body temperature is also kept low by evaporating the perspiration during hot periods.

(c) Viscosity

The viscosity of water is high because of the energy contained in the hydrogen bonds. Viscosity of water can be visualized best if one observes it or any other liquid flowing through a glass tube or a clear plastic hose. The liquid moving through the tube behaves as if it consisted of a series of parallel concentric layers flowing one over another. The rate of flow is greater at the centre, but because of the amount of internal fraction between layers, the flow decreases towards the sides of the tube. This phenomenon can be observed along the side of any stream or river with uniform banks. The water along the banks is nearly still, while the current in the centre may be swift. This resistance between the layers is called *lateral* or *laminar viscosity*. In another kind of viscosity, called *eddy viscosity*, water masses pass from one layer to another and creates turbulence both horizontally and vertically.

Viscosity is the source of frictional resistance to objects including organisms moving through water. Since this resistance is 100 times that of air, animals must expand considerable muscular energy to move through water. A mucous coating on fish, *frogs* and certain other aquatic organisms reduces surface tension. Streamlining of body does likewise. In fact the body forms of some aquatic organisms have evolved under the stresses of viscosity. The faster an aquatic organism moves through the water, the greater the stress placed on the surface and greater the volume of water that must be replaced in a given time. Replacement of water in the space left behind by the moving animal adds additional drag on the body. An animal streamlined in reversed direction, with a short, rounded front and a tapering body and meets least resistance in water.

The viscosity of water allows organisms to swim using relatively simple movements. It also protects the aquatic animals and plants from the mechanical disturbances.

(d) Surface tension

Water particles are attracted to one another. Molecules of water below the surface are symmetrically surrounded by other molecules. The forces of attraction are the same on one side of the molecule, as on the other side. But at the water surface the molecules exist under a different set of conditions. Below is a hemisphere of strongly attractive similar water molecules; above is the much smaller attractive force of the air. Since the molecules on the surface are drawn into the liquid, the liquid surface tends to be as small as possible, taut-like the rubber of an inflated balloon. This is surface tension, which is greatest among all common liquids, except mercury.

(e) Cohesion and adhesion

Cohesion is the ability of water to stick itself, and adhesion is the ability of water to stick to some solid substance. Both these properties can be explained on the basis of the presence of hydrogen bonds.

The property of cohesion gives water a very high surface tension which is caused by the strong attraction of polar water molecules for each other by strong hydrogen bonds on the surface of water. Cohesion is demonstrated by the formation of droplets of water. Several insects such as water spiders are able to move on the top of water without breaking the surface film because of the surface tension of water.

The capillary action demonstrated by the rise of water in a glass capillary occurs only due to these properties of water. Water adheres to the glass because the hydrogen in the water molecule is attracted to the oxygen in the glass without breaking its internal cohesion. Water in the soil can rise several times by capillary action due to adherence to soil particles where oxygen atoms of silica attract the hydrogen atoms of water molecules.

(f) Solvency

No other compound can be compared to water as a *solvent.* So many different substances can be dissolved in it that it is known as the *universal solvent.* More things, in fact, can be dissolved in water than any other liquid. This is especially true for inorganic chemicals which split, or dissociate to form electrically charged entities termed ions. Ionization influences most electrical phenomena and many chemical phenomena of solutions. It is probable that all natural elements are soluble in water, atleast in trace amounts, and that they are found in natural water at some place or other on the earth's surface. Thus, water is the main medium by which chemical constituents are transported from one part of the ecosystem to the other. It is the only medium by which these constituents can pass from the abiotic portion of the ecosystem into the organisms. Even in the arid terrestrial environments, nutrients pass into the roots of plants in aquous solution; when air is breathed by animals, oxygen is dissolved in water at the surface of the lunfs before it can cross the mucous membrane and be absorbed by the blood.

(g) Density

Ice is approximately 9 percent less dense than water. The maximum density of water at atmospheric pressure is at 4°C therefore, ice or water at any other temperature will float on it.

This property of water is mainly due to hydrogen bonds. In ice, the molecules become arranged in open hexagonal rings with large intermolecular spaces having four hydrogen bonds per water molecule. As the ice melts into liquid phase, several bonds are broken. The molecules get more compacted and the water becomes more denser. There are average 2-3 bonds per molecule in liquid water. At 4°C, the compactness in the arrangement of water molecules is maximum. With the rise of temperature, the molecules begin to get apart, occupying more and more space until at 100 °C when the water boils and the liquid is converted into gaseous phase.

The density of water cause the temperature stratification, and vertical mixing of lakes which are helpful in many ways to the aquatic organisms. The complete freezing of water bodies is also checked. In cold seasons when the surface water approaches 4°C, it sinks to the bottom, so only surface of lakes in freezed while the bottom temperature remains at 4°C keeping the organisms alive.

(h) Transparency

Water is transparent medium. Its transparency enables the penetration of light to the depths where it is ultimately absorbed. Water absorbs light/ transforming radiant energy into heat. But it does not absorb all wavelengths equally (Figure 8). Both ends of the visible light spectrum, especially the red-end are selectively absorbed, so that as one descends into a body of water the colour changes from white to bluish to a dull blue-green. Light may be diffused or absorbed by particles in water such as sediments, detritus, animals and plants. The more particles in water, the greater the degree of diffusion and absorption. Figure 9 illustrates the amount of light transmission in different bodies of water under various conditions of turbidity. The zone upto which light rays penetrate is called *photic zone* and below this zone is complete darkness.

The most significant abiotic effect of light penetration is heating of water. However, the most important biotic function of light is its role in photosynthesis. In contrast to the air of terrestrial systems, light is absorbed by water very much faster, so light is an important limiting factor in an aquatic ecosystem. In order to

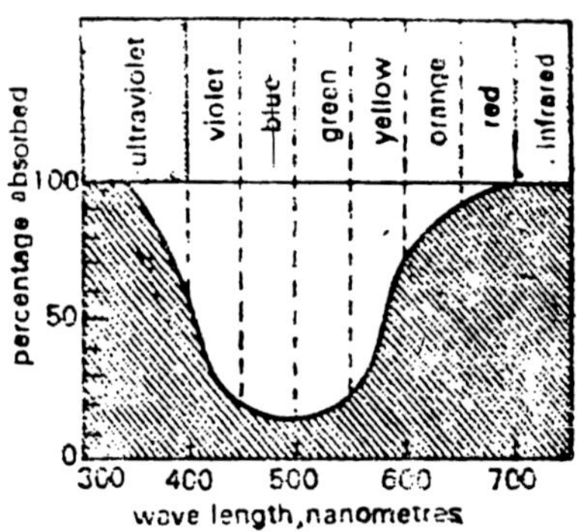

Fig. 8. Absorption of different wave lengths of light by 10 metres of pure water (after Clapham, Jr., 1973).

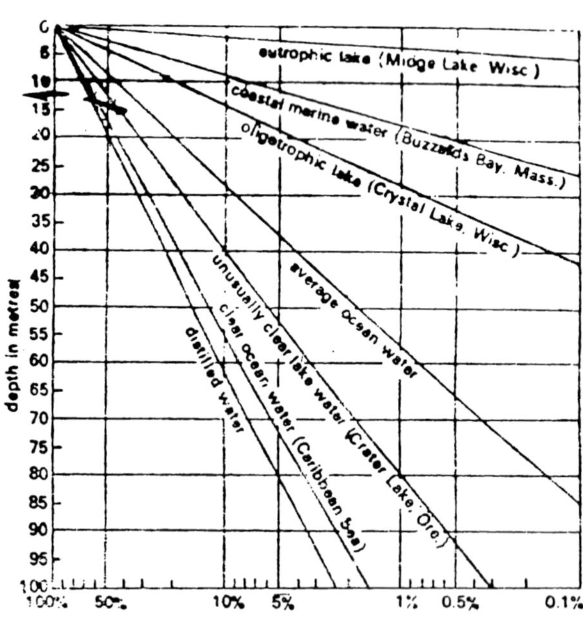

Fig. 9. Rate of absorption of light in different bodies of water. The curves are based on the yellow-green (500-600 nm) portion of the spectrum (after Clapham, Jr., 1973).

survive, a plant must must fix more energy that it utilizes in its respiration. That is the ratio between gross primary production and respiration must be greater than one. There is a level in water at which the mean daily light flux allows a rate of production equal to that of respiration. This flux is termed the *light compensation level.* Above the compensation level, autotrophs can produce sufficient food for themselves and for the animals associated. Below the compensation level, all energy must be derived from organisms living and feeding above the compensation level and sinking below it, either as part of their normal behaviour or after death. The compensation level fluctuates under various circumstances. It tends to be higher in winter than in summer because the total radiation flux is lowest in the winter. In highly turbid areas, due to light scattering and great diffusion of light, the compensation level raises upto surface of water.

(i) Pressure

Organisms living at sea level experience a pressure of about 15 psi, which is defined as 1 atm (760 mm of mercury). Pressures increase with increased depth of water at the rate of 1 atmosphere for every 10 meters of descent. Organisms inhabiting the floor of deep sea areas at depths of 10,500 meters are exposed to hydrostatic pressure of about 1 ton per square centimeter. Pressure influences solubility, ionic dissociation, and surface tension, and water is slightly compressible with increased pressures.

(j) Salinity

Salinity has been defined as "the total amount of solid material in grams contained in one kilogram of water, when all the carbonate has been converted into oxide, bromine and iodine replaced by chlorine and all organic matter completely oxidised". All types of natural waters contain various amounts of different salts (ions) such as sodium, potassium, magnesium, chlorine, sulphate, phosphate, carbonate, bicarbonate, nitrate etc., and all these salts are responsible for the saltiness, salinity or salt content of water. The salinity of marine water is rather constant being about 3.5 percent. The salinity of fresh water varies greatly (Table 5). Some salt lakes may have a salinity of 25 - 30 percent which greately restricts life in them.

Table 5. Comparison of some of the principal ions found in different kinds of water.

Water	Na	K	Ca	Mg	Cl	SO_4	CO_2	Total/ litre
1. Soft water	0-016	—	0-01	00005	0-019	0-007	10-012	0-065
2. Hard fresh	0021	0.016	0.62	0014	0.041	0.025	0.116	0.331
3. Sea	10.5	0.30	0.40	127	18.98	2.65	0.71	34.85
4. Great salt lake	65.54	3.76	0.065	4.47	10.08	13-04	—	197.51

CHEMICAL PROPERTIES OF WATER

Water consists primarily of a single compound, H_2O. It is a universal solvent and most chemical compounds ionize readily in water and provide many radicals and considerable versatility in the rearrangement of chemical substances. It has following chemical properties.

Solubility of Gases in Water

Most gases dissolve readily in water, most notably those that are essential for life. The concentration of any gas in water genrally varies between zero and a theoretical maximum or saturation. The latter is the amount of gas that can be dissolved in water when the atmosphere and the water are in equilibrium with one another. Except for water falls and very turbulent streams, the water in natural ecosystem is seldom in equilibrium with the atmosphere. A gas may shew a deficit, if it is being utilized in the ecosystem faster than it is going into solution across the air water interface, or it may be supersaturated, if it is being produced in the ecosystem faster than it is being released from solution across the interface. The concentration of important gases may vary widely in any ecosystem, both horizontally and vertically.

The saturation level of any gas in water defends on several variables, most notably temperature, salinity, the concentration of the gas in the atmosphere, and its relative solubitiiy in water. The greater ihe concentration of a gas in the atmosphere, the greater its concentration in the water will tend to be. depending on its relative solubility in water.

Oxygen. One of the most critical factor in an aquatic environment is the amount of oxygen in the water because most living organisms (excepting anaerobic forms) require this gas for respiration. In contrast to atmosphere, the oxygen becomes limiting factor for aquatic animals as the saturation concentration of oxygen in water is governed by temperature and salinity. As is evident in table 6, the lower the temperature, the greater the oxygen retaining capacity of water, whether it is fresh-water or sea water (Fig. 10)

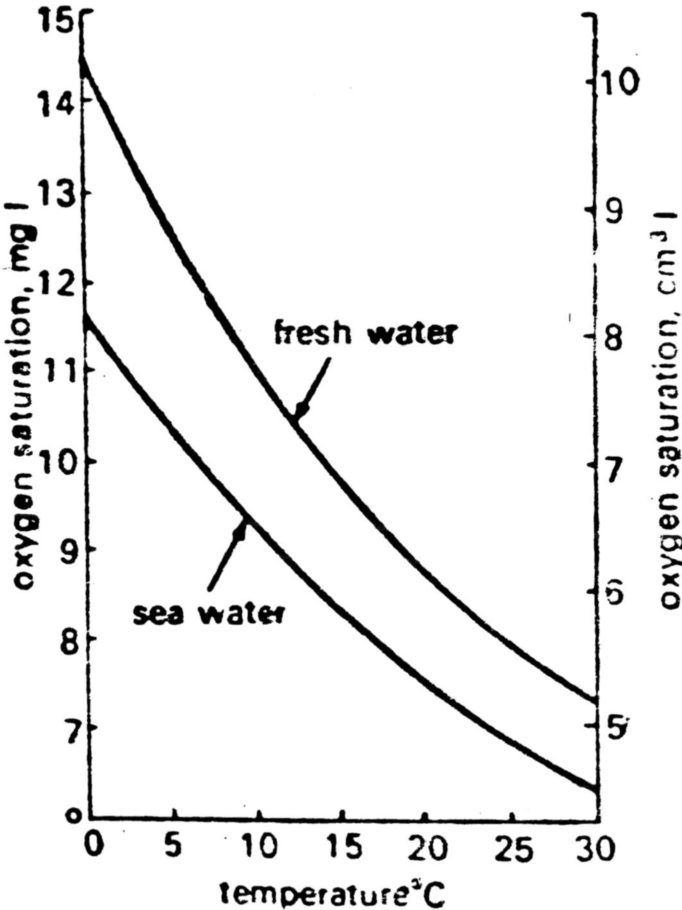

Fig. 10. Curves for oxygen saturation with respect to temperature in fresh and sea water

In fresh-water areas of any depth and in the suit water habitat, there are three recognizable zones with regard to oxygen concentration. A surface stratum, where the oxygen tends to be in equilibrium with the atmosphere above. Below this su· face stratum of variable depth is an intermediate stratum where oxygen values fluctuate in accordance with existing factors. Respiration, decomposition of organic materials (stagnant ponds) and stream pollution all tend to reduce the amount of available oxygen, while photosynthetic activity will often balance or more than balanace oxygen loss. The deepest layers of water will usually have a very low oxygen concentration in the deeper lakes and oceanic areas because the continual decomposition of organic debris, the respiration of organisms inhabiting these deeper water, and the complete absence of photosynthetic activity in these lower strata will tend to deplete the oxygen concentration. The deep stratum is entirely dependent upon the slow transport of oxygen from the overlying intermediate layer.

Table 6. Comparison between equilibrium concentrations of important gase in the atmosphere and in water (Clapham, Jr., 1973).

Gas	Atmosphereic concentration	Saturation in water
Oxygen	210 cc./1.(21%)	7cc./1 (32.9%)
Nitrogen	780 cc./(1.078%)	14 cc./1(65.7%)
Carbon dioxide	0.3 cc./1(0.03%)	0.3 cc./1(1.4%)

Because oxygen is needed for the respiration of all heterotrophs, it tends to be taken out of solution continuously by the organisms. The only means by which it can enter the water are via solution at the air-water interface or through photosynthesis by aquatic plants. Oxygen is transferred to deeper water by diffusion or through circulation of the water. The rate of oxygen solution at the surface is highly variable, depending mainly on the turbulence of the water and the presence of waves spray and foam.

Table 7. Comparison of the saturation concentration of oxygen in freshwater and salt water environment with varying temperatures.

Water	Temperature °C	Saturation point in millilitre per litre
Fresh water	0	10.27
Salt water	0	8.08
Fresh water	30	5.57
Salt water	30	4.52

Nitrogen. Nitrogen is significantly less soluble in water than oxygen. But because it constitutes 78 per cent of the atmosphere, it still accounts for about 65 per cent of the dissolved gases at equilibrium. It is fairly inert chemically and does not react with water, although some bacteria, fungi, blue-green algae, and so on, can use it to satisfy their nitrogen requirements, and other bacteria can produce it througth reduction of nitrate under conditions of very low oxygen concentration.

Carbon dioxide. The decomposition of organic matter and the respiratory activity of aquatic plants and animals produce carbon dioxide. This gas is one of the essential raw materials necessary for photosynthetic activity by green plants. Carbon dioxide combines chemically with water to produce carbonic acid (H_2CO_3) which influences the hydrogen ion concentration (pH) of water. Carbonic acid dissociates to produce hydrogen (H^+) and bicarbonate (HCO_3^-) ions. The bicarbonate radical may undergo further dissociation forming more hydrogen (H^+) and carbonate (CO_3^-), as represented in following reactions:

$$\underset{\text{atmospheric}}{CO_2 + H_2O} \underset{\text{air-water interface}}{\rightleftarrows} \underset{\text{dissolved}}{CO_2 + H_2O} \rightleftharpoons \underset{\text{carbonic acid}}{H_2CO_3} \rightleftharpoons \underset{\text{bicarbonate}}{H^+ + HCO_3^-}$$

$$\updownarrow$$

$$\underset{\text{carbonates}}{2H^+ + CO_3^-}$$

The amount of free or uncombined carbon dioxide in wafer is of ecological importance : it governs the precipitation of calcium in the form of calcium carbonate ($CaCO_3$). Calcium precipitates

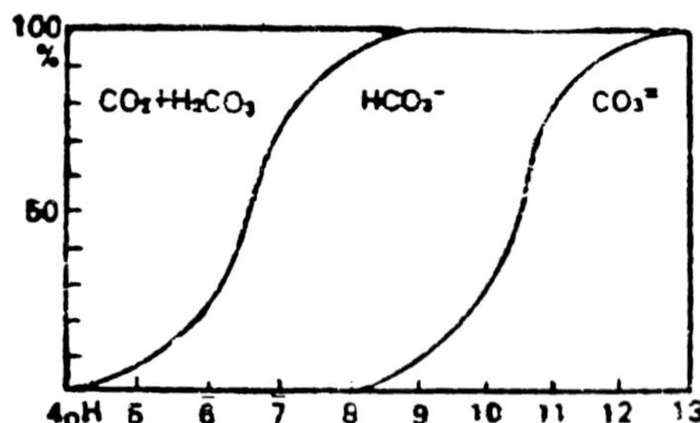

Fig. 11. Proportions of various forms of carbon dioxide over the ranee of pH commonly found in ecosystems.

when temperatures and salinity are high and the amount of uncombined carbon dioxide is low. This means more carbonate (CO_3^-) is present to combine with the calcium cation (Ca^{++}). These conditions exist in shallow tropical waters, where evaporation is high. This raises the salinity and photosynthetic activity of plants and reduces the quantity of free carbon dioxide in water. The precipitation of calcium carbonate in tropical areas as the Bahamas explains the preponderance of thick calcareous shells of shallow water tropical molluscs, plankton and algae. In deep oceanic water, temperatures are low and there are no photosynthetic plants, consequently, the carbon dioxide content of the water is high. Deep water fauna (molluscs, crustaceans) possess very fragile skeleton because the precipitation of calcium carbonate is minimum.

Hydrogen sulphide. The deeper layers of many bodies of water, including ponds, lakes, and some estuaries, may contain significant amounts of the toxic gas, hydrogen sulphide, which is released by decaying organic matter. If concentration of the gas builds up, all life but anaerobic bacteria excluded from the area (*e.g.*, deeper strata of Black sea).

Hydrogen-ion Concentration (pH)

One property of natural water is their acidity or alkalinity. About one water molecule in every 10,000,000 splits into two, producing a hydrogen ion (H^+) and a hydroxyl ion (OH^-) :

$H_2O = H^+ + OH^-$

In pure water, there are equal numbers of H^+ and OH^- ions, it therefore, has a neutral reaction. Some, natural waters, however, acquire an excess of H^+ and are acidic, while others, with an excess of OH - are alkaline. For example, in fresh.-water habitats such as acid bogs, swamps and drainage creams carrying water from these areas are acidic, contain pH value as low as 1.4. Despite the high acidity, rich populations of acidophilic flora and fauna thrive under such conditions. On the other hand, certain lake waters may be quite alkaline, particulraly in lime stone areas, where the pH may range from 10 to 12. Basophilic plants and animals are found in these areas. In a vast majority of fresh-water localities pH value ranges between 5.5 and 8.5. Oceanic areas exhibit little change in pH values over vast areas because of the effective buffering action of anions such as HCO_3^- and CO_3^-. Surface waters of open ocean have pH values of 8.0 to 8.4, but deeper water is close to neutrality (pH 7.4 to 7.9). There is often a wider range of pH values in shallow marine water, estuaries and tide pools.

HARD AND FRESH-WATERS

Water containing soluble salts of calcium and magnesium such as chlorides, sulphates and bicarbonates, is called hard water. The hardness of water may be of two types—temporary hardness and permanent hardness. Temporary hardness is caused by dissolved calcium bicarbonate or magnesium bicarbonate. Temporary hard water can be converted into soft water by boiling; as at higher temperature soluble bicarbonates of calcium and magnesium are converted into insoluble carbonates of calcium and magnesium. Permanent hardness of water is caused by dissolved chlorides or sulphates of calcium and magnesium. This hardness cannot be removed by boiling and requires some specific chemical treatments.

USES OF WATER

Water use has been considered in two categories: (i) non-consumptive use that remove water from its natural courses and (ii) consumptive part of it. On a national basis, water for agricultural purpose is equal to 80 percent of the total national use, that for domestic purpose is equal to 4 percent, and for industries it is only 16 percent. In agriculture sector it is 67 and 33 percent for cunsumptive and non-consumptive use respectively, In industrial sector it is 8 and 92 percent in consumptive and non-consumptive use respectively. While, it is 10 and 90 percent in domestic sector for consumptive and non-consumptive use respectively (Figure 12)

The consumptive use of water is the amount of water that is taken out of a stream or pumped out of an underground or surface reservoir in order to reach its point of use. The major use of water in this category is in public supply (domestic consumption, irrigation, livestock and industry). The total water demand estimated by 2000 AD in India for various purposes have been shown in Table 8.

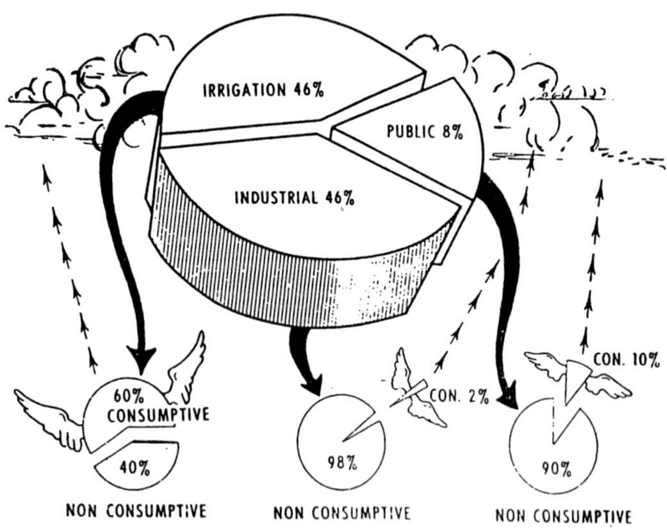

Figure 12. Water use : Consomptive and non-consumptive.

Table 8. Estimated water demand in India by 2000 AD in thousand million cubic meters (Rao, 1979)

Water use	% Used	Water used	Water consumed	Water reused
Agriculture				
(a) Irrigation	77	860	774	86
(b) Livestock	1	9	9	–
Power	13	150	5 1	45
Industries	3	35	10	25
Municipal and rural water	6	62	31	31
TOTAL	100	1116	829	287

In case of non-consumptive use, water is used without being removed from where it is naturally present, such as for navigation and transport, swimming, boating and other recreational purposes, wildlife habitat, aquaculture and for waste dilution. The use of water in generation of hydroelectric power is also a non-consumptive use, since water remains in the system. The dumping of domestic wastes and industrial effluents is the single greatest non-consumptive use of water which significantly affects the quality of water.

Our natural water resources have, always been subjected to conflicting uses. We are all different in some way, yet each of us has a fair-share heed of water. Many of the priorities which we set for use of that fair-share compete and conflict with priorities set by others. Observations (Table 9) illustrates how water is being used among the three major types of water uses and how this use is expected to grow by 2000-2001 if our anticipated demands are realized.

Table 9. Estimated water use and projected requirements in Billion litres per day (Chaturvedi, 1975).

Use	Type	1973-74	1978-79	1988-89	2000-01
Domestic	NC	34.20	49.78	60.50	85.00
	C	17.42	21.95	30.90	43.50
Industrial	NC	12.30	18.90	47.20	151.00
	C	0.99	1.49	358	11.58
Agricultural	NC	1333.15	1718.23	2562.70	4093.00
	C	807.76	1038.79	1531.20	2380.38

NC = Non Consumptive use, C = Consumptive use.

In another estimated requirement for freshwater for different uses for the years 2000 and 2025 (Table 10) indicates the demand of water in the coming decades and therefore the need to maintain water resources to ensure adequate freshwater supplies for various uses.

Table 10. Estimated requirement of freshwater for various uses in India (Nag and Kathpali,1975).

	(Values in $10^9 m^3$)	
Use	2000	2025
Irrigation	630	770
Domestic & livestock	34	50
Industry	30	120-150
Thermal Power	60	160

Table 11. Estimated requirement of freshwater for various uses in India for the year 2000 (Rao, 1975).

	(Values in $10^9 m^3$)		
	To draw	To consume	To return
Irrigation & livestock	869	783	86
Power	150	5	145
Industry	35	10	25
Domestic (urban)	38	8	30
	1092	806	286

The availability of water has been estimated to be 1900 x 10^9 m^5

(a) Domestic use

Water requirement for drinking and sanitary purposes is not large but it is the most essential use. It is extremely important from health, convenience and efficiency point of view. The requirement of clean water per person is about 2.7 litres per day. The average person drinks or otherwide uses about 70,000 litres of water during his life time. In the United Kingdom the demand for domestic water by 2000 AD was about 235 litres per person per day. However in the warmer regions of the world, the domestic demand might have gone upto 500 litres. In India, the water consumption was 90 litres per person per day in the rural areas and 270 litres in urban areas. Thus the minimal amount of drinking water at the global level needed annually is about 5 billion cubic meters. In view of our growing population, a colossal effort will be required to supply this basic seed (Chaturvedi, 1975). Drinking water supplies are mostly available in urban areas. However, millions in rural areas do not have drinking water supplies of standard quality mainly due to other investment priorities (Lahiri, 1975).

The supplies for domestic use should be pure and wholesome. Wholesome water is that which could be consumed without risk from its bacterial or chemical content. Water from a properly selected source, and submitted to an adequate system of purification, can be called pure and wholesome, if it is free from visible suspended matter, colour, odour and taste, and all objectionable bacteria indicative of disease producing organisms. It should also not contain dissolved matter, either mineral or organic, which in quality and quantity would render the water dangerous to health.

(a) Agricultural Use

The total quantity of water used for irrigation is largest. It has been estimated that nearly 3500 litres of water per person per day is used for irrigation. This quantity is several times higher than the average domestic demand. Estimates show that for obtaining 1 kg each of wheat, rice, meat and milk about 600, 2000, 25000, and 400 litres of water is required respectively.

One of the facts is that largest uses of water are in agriculture. This is primarily in view of the agrarian economy of our country. Further, well in the twentyfirst century, on the past estimates of water availability and use, even with the withdrawl of all available resources, these will be short of our agricultural demands.

(c) Industrial Use

Industry of all kinds is a minor user of our water. The importance of water to meet industrial needs have been shown in Table 12, which indicates the amount of water needed to produce various amounts of major industrial commodities.

Table 12. The estimated industrisl water use for a few typical industries in India (Sundereshan and Subrahmanyam, 1980).

Industry	Production	Water requirement (1/unit product)
Dairy	Milk, (1)	6-10
Distillery	Alcohol, (1)	20
Cotton Textile	Cloth, (m)	10-70
Viscose Royon	Fibre, (kg)	1600
Tannery	Hide, (kg)	40-45
Coke Oven	Cokes, (kg)	*
Phenol formaldehyde resin	Resin, (kg)	*
Oil Refinery	Oil, (kg)	*
Pulp and Paper	Paper, (kg)	270-450
Sugar	Sugar, (kg)	15-40
Integrated steel mill	Steel, (kg)	20-50
Urea	Urea, (kg)	6-8

* Data not available.

Less than 10 percent of the industrial units, including power generation, account for 80 percent of the total industrial intake. Excluding power stations - thermal and hydal both -the major users are paper industries, the cotton and jute industries, sugar industry, fertilizer industry, steel industry and oil refineries.

Industry uses more water than any other commodity. according to estimates more than 630 billion litres of water is consumed per day by industry in United States. One ton of steel making consumes about 3,00,000 litres of water. For making an average automobile, about 4,50,000 litres of water is needed, Most of the water used by industry is for cooling.

For example, water cools the hot gases produced by refining oil and hot steel in steel mills. In an estimate by UNO (Table 13) aluminium smelting uses maximum water (1350 litres/kg of product) while baking industry uses minimum water (3 litres per kg of product).

Table 13. Water demand in industry as estimated by UNO 1958 (Tebbutt, 1977).

Industry or product	Typical water demand (L/kg of product)
Aluminium smelting	1350
Baking	3
Brewing	15
Butter	11
Coal mining	4
Cheese	20
Fruit canning	20
Glass	70
Laundering	30
Oil refining	10
Paper	160
Steel	45
Sugar	9
Sulphuric acid	10
Synthetic fibres	200
Vegetable canning	10
Wallboard	100

Chapter-2

WATER POLLUTION

Water pollution is emerging as a threat to all mankind. The dirty world of water pollution is broad and complex, and just getting a grasp on it, so that it can be discussed intelligently, presents problems. We are different in some way, yet each of us has a need for our "fair share" of water. Since we are all different, many of our priorities which we set for use of that fair share compete and conflict with the priorities set by others. As a result, the mental images which each of us forms, when he thinks of water pollution rarely match with those conceived by others. When compared in discussion with others, they produce, in fact, results much like those which might stem from several blind persons examining, then describing an elephant.

The term "water quality" is intimately related to water pollution. Polluted water is that water which has more "negative" qualities than it has "positive" ones. Water quality refers to the physical, chemical and biological characteristics of water. The physical characteristics include the temperature, clarity and similar qualities. Chemical water characteristics include the presence, amount if present, organic and inorganic substances in solution, and the way that these substances in solution are bounded or dispersed in water. Biological characteristics of water include identity, impact and organisms which are present in water. In simple words, polluted water is water which has been abused, defied in some way, so that it is no longer fit for some use.

The physical, chemical and biological characteristics of water are formed not only during its penetration through the atmosphere, soil and rocks but also during its contacts with the vegetation canopy and cultivation practices. Water from afforested areas are generally of good quality. The replacement of forest poluculture by monocultures leads to an increase in acidity. Water from afforested catchments can also become bacteriologically contaminated by wildlife, and particularly by birds. The pollution of water resources is a consequence of:

— Natural processes: erosion, volcanic activities, biological processes and human activities.

— Increasing erosion owing to deforestation, wrong cultivation practices and urbanisation.

— Washing of agrochemicals from agricultural and silvicultural production.

— Accident during the transport of fuel and other chemicals.

— Disposal of gaseous, liquid and solid washes from industry, thermal and nuclear power generation, agriculture, dwelling areas etc.

— Subsequent leaching of wastes deposited on the surface, under the ground, or in water.

— Infiltration of polluted water from terrestrial sources to the ground water resources (Jemmer, 1987).

We are facing the pollution caused by Municipal sewage, domestic and industrial use of synthetic detergents, industrial wastes and agrochemical run-off from the fields (Figure 13). According to the survey carried out by several workers on selected stretches of some of the important rivers, it has been observed that the water of most of our rivers is polluted (Agarwal, 1990). During the last few years much concern has developed over the continued deterioration of our lakes. This is mainly due to uncontrolled human activities which are responsible for the accelerated flow of minerals from the terrestrial to the aquatic portion of the watershed (Vyas et al, 1982).

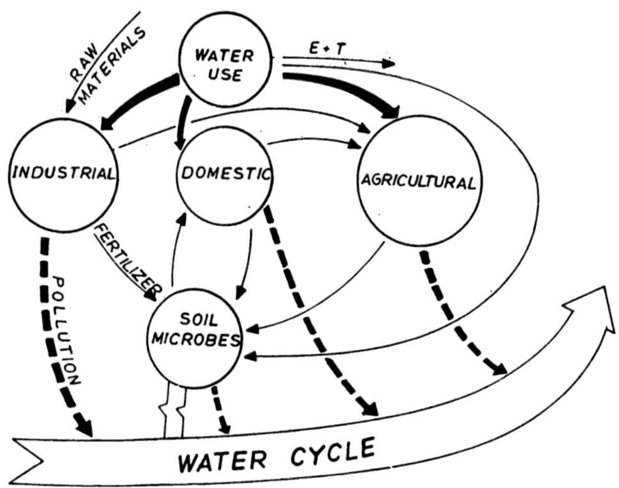

Fig. 13. Sources of water quality deterioration.

The ancient civilization has perceived the relations between water quality and human health and every effort was made to prevent the pollution of drinking water from contamination by human wastes. As long as, the human population was small and communities were scattered over large areas of land, the waste disposal created no problems. It was left to nature to dispose it by assimilation in the surrounding air and land. But with increasing population especially on the banks of water bodies, the soiled water called sewage, was channelized to streams and rivers. At the beginning, this mode of disposal was considered to be quite suitable. But with rapid urbanisation the natural waters have been polluted to such an extent that they have become unsuitable as sources of water supply (Alone, 1988). In towns of India, there are no facilities for efficent sewage disposal system.

Our natural water resources since decades have been subjected to different levels of ecological stress, by increasing pace of industrialization in public and private sectors. The effluent is an inevitable consequence of industrial process (Singh et al, 1985).

The discharge of effluents into surface water has exceeded the assimilative capacity of the receiving streams at a number of places (Jain and Agarwal, 1989), and is receiving greater dimensions day-by-day. So long as the pollutional stresses are within elastic limits, the ecological systems recover and tend to come to the original state. But once, the elastic limit is crossed, irrevocable stresses are set and the distorted ecological equilibrium does not return to the original condition observed under no stress condition (Shrotriya, 1984).

The nature and quantity of wastes differ from industry to industry, depending upon the nature of raw materials, processes used, products manufactured and by-products recovered. Among all refining, mixing, blending, extraction and manufacturing operations some wastes are produced. Besides accidental spills, leakages and so forth, which may occur inspite of our best efforts and leads to grave consequences.

WATER POLLUTION DEFINITIONS

Water pollution can be defined in a number of ways:

1. Water pollution is the presence of any foreign substance (organic, inorganic, radiological or biological) in water which tends to degrade the quality so as to constitute a hazard, or impairs the usefulness of water (USPHS Drinking Water Standards, Item 6, 1962, p.1).

2. Water pollution include "discharges of wastes" including radioactive or other substances actually or potentially harmful to such uses and alteration of properties of water in such a way as to be harmful, including temperature, taste, turbidity or odour (FWPCA, USA).

3. Courts of law define water pollution as "any impairment of water quality that makes water unsuitable for beneficial use" (F.E.Moss, the water crises, New York, 1967,p.64).

4. Water pollution is adding to water of any substance or the changing of water's physical characteristics in any way which interferes with its use for any legitimate purpose (US Senate Select Committee on National Water Resources,

Committee print No 9, January 1960, p.1).

All these definitions show some similarities and some dissimilarities; however, the basic elements of most definitions are the concentrations of particular pollutants in water for sufficient periods of time to cause certain effects. If the effects are health related, such as those caused by pathogenic bacterial intrusion, the term "contamination" is appropriate. Effects that have to do with limitations on water availability due to certain water quality requirements related to usage can serve as a basis for defining a condition of water pollution. "Nuisance" refers to aesthetically displeasing effects created by oils, grease, or other floating materials (Canter, 1977).

SOURCES OF WATER POLLUTION

There are two main sources of water pollution in a water body: (i) point sources, and (ii) non-point sources. The total waste load in a water body is represented by the sum of all point and non-point pollutant sources (Shubinski and Tierney, 1973).

Individual sources of water pollution are manifold, and, resultant water quality probably will not be attributable to single type of pollutant or single source. Some of the major sources of water pollution are municipal waste, industrial wastes, agricultural wastes, erosion, accidental spillage of oil and other hazardous substances, acid mine drainage, mine sediments and watercraft waste (Federal Water Quality Administration, 1970).

Pollution of water resources can be caused by one or more of the following sources: (i) atmospheric dissolved gases, (ii) weathering of soil and rock minerals, (iii) decomposition of animal and vegetable materials, and (iv) industrial effluents, sewage and municipal wastes.

(a) Atmospheric gases

When the concentration of dissolved gases viz., Nitrogen, Oxygen, Hydrogen, Carbon dioxide, Ammonia, Sulphur dioxide, Hydrogen sulphide etc. increases sufficiently in the atmosphere, it subsequently enhances their concentration in water of their contact. However, the actual concentration of these gases in water depends upon their concentration in the atmosphere and their solubility and diffusivity in water, temperature and viscosity of the water. Generally surface waters near the cities and industrial complexes where the atmosphere is more polluted by industrial emissions and automobile exhaust are more polluted than those of rural areas. Moreover, stagnant waters like those of lakes and ponds are more polluted than the flowing waters of rivers and streams.

The concentration of gases in water in contact with the atmosphere depends upon the degree of equilibria being maintained between the water and the gaseous phases. The rate of absorption of gases in, and extraction of gases from the polluted waters depends upon the diffusion coefficient of the gases in a liquid-gas film at the gas-water interface.

The diffusion of gases into, and from water is further complicated by the presence of surface active materials like detergents in polluted waters, These are known to concentrate at the gas-water interface and are likely to affect the diffusion coefficient values for gases in the gas-water film. The gaseous composition of moisture free air in 78.08 percent Nitrogen, 20.95 percent Oxygen and the rest other gases. Of the common atmospheric gases, Nitrogen, Oxygen, Hydrogen and Methane do not react appreciably with water. Hence, their solubility in water is directly proportional to their partial pressure at equilibrium with the atmosphere. However, the solubility of gases such as Hydrogen,sulphide, Carbon dioxide, Sulphur dioxide and Ammonia, though found in much less quantities, are much higher (Table 14). The main source of Oxygen in natural water are air and chlorophyll bearing aquatic plants and algae.

Table 14 . Solubility of gases (ml/L) in pure water in contact with gas at pressure of 1 Atm.

Temp. °C	N_2[1]	O_2[1]	H_2[1]	CH_4[1]	CO_2[1]	H_2S*	SO_2*1	NH_3*2
0	29.40	69.8	1.93	39.8	3,360	7.100	186,700	471,000
5	26.10	61.2	1.84	34.3	2,790	6,040	162,300	438,000
10	23.20	54.3	1.76	29.9	2,345	5,160	140,600	406,000
15	21.10	48.7	1.70	26.4	2,000	4,475	170,500	375,000
20	19.30	44.3	1.64	23.6	1,720	3,925	103,300	344,000
25	17.90	40.4	1.58	21.5	1,495	3,470	88,200	322,000
30	16.80	37.2	1.53	19.7	1,305	3,090	74,700	288,000
40	14.80	32.9	1.48	16.9	1,040	2,520	54.800	252,000
60	12.77	27.8	1.44	14.0	704	1,810	-	192,000
80	11.96	25.1	1.44	12.7	-	1,394	-	134,000
100	11.35	24.2	1.44	12.2	-	1,230	-	-

1 Source: Landolt-Bornstein, Physikalische *Chemische Tubellen,* Berlin, Springer, Verlag, 1923, p. 763.

2 Source : Landolt-Bornstein, ibid., also J. W. Mellor. *Comprehensive Treatise on Inorganic and Theoretical Chemistry,* vol. 8, London Longmans, 1922, p. 194.

* Total solubility including ionization products.

The concentration of Oxygen in natural water determines the extent and rate of pollution. As a matter of fact, Oxygen content, directly or indirectly, controls the concentration, solubility and diffusivity of other gases like Hydrogen sulphide, Carbon dioxide. Ammonia and Sulphur dioxide etc., and a sort of chain reaction continues in the water body. For example, dissolved Hydrogen sulphide cannot exist in the presence of oxygen, but will be oxidised to water and Sulphur according to the following reactions:

$$2H_2S + O_2 + 2\,H_2O + 2\,S$$

$$H_2S + 2\,O_2 = 2\,H^+ + SO_4^-$$

Hydrogen sulphide gas is produced by the anaerobic decomposition of materials, and also by the dilution of waste sulphate liquors discharged from textile mills, pulp and paper mills and tanneries. Since, primarily Hydrogen sulphide is not a constituent of atmosphere, and secondly, it is also easily oxidised by oxygen, it can only be liberated by aeration. However, the rate of Hydrogen sulphide liberation will depend upon the rate of agitation. Hydrogen sulphide can also be removed by oxidising with chlorine (Hack and Goodson, 1952).

$$H_2S + Cl_2(aqu) = 2\,H + 2\,Cl^- + S$$

Sulphur dioxide gas enters natural waters from combustion of fumes in the air and oxidation product of sulphur-bearing organic matter. Sulphurous and sulphuric acids are present in waters of coal mines. Its solubility in water is quite high and almost completely above pH 4 but it cannot exist in the dissolved oxygen because of its oxidation. However, SO_4 can be reduced to H_2S under highly anaerobic conditions.

$$SO_2 + H_2S = H^- + HSO^-_3$$

$$HSO_3 + \tfrac{1}{2}\,O_2 = H^+ + SO^-_4$$

Nitrogenous organic matter, on bacterial decomposition, both in anaerobic and aerobic conditions, produces Ammonia, and it is the main nitrogenous food of the organisms. It may be realeased by the industry as a waste product or in sewage along with unhydrolysed urea from the urine. On hydrolysis, urea gives Ammonia and Carbon dioxide. Ammonia is highly soluble and with water it ionises (Table 15) below pH 6, and is non-ionised above pH 11. Presence of small amount of free Ammonia (0.2 ppm) in natural waters is only an indication of recent organic pollution, and in highly saline (or sea) water Ammonia concentration ranges upto 0.075 ppm (Hood, 1966).

Table 15. Ionisation (%) of Ammonia in dilute solution under different pH at 25°C.

pH	NH_4+(%)	NH_3(%)
5	99.99	0.005
6	99.95	0.05
7	99.45	0.55
8	94.77	5.23
9	64.50	35.50
10	15.30	84.70
11	1.78	98.22

Carbon dioxide concentration in the atmosphere is about 0.03 percent by volume, but is generally more in the vicinity of industrial areas. It is also the decomposition product of organic matter both in anaerobic and aerobic conditions. It is also contributed by respiration of organisms. On reacting with water it ionises as follows:

$$CO_2 + H_2O = H^+ + HCO^-_3$$

At high pH it is further ionised to

$$HCO^-_3 = CO^-_3 + H^+$$

Its solubility in fresh water is 0.03 percent. Natural waters are generally super-saturated with Carbon dioxide, particularly in summer when the rate of decomposition of organic matter is high. However, solubility of Carbon dioxide decreases with the increase of salinity, and the reduction is about 3 and 7 percent at 5 and 10 thousand ppm salt concen-tration respectively (Fox, 1909).

Methane is not a permanent constituent of the atmosphere while Hydrogen is found to the extent of 0.001 percent of the earth's atmosphere. The presence of these gases is primarily due to anaecobic decomposition of organic matter by bacteria.

Methane has been found quite high in swamp waters and coal mine drainage, and more than 20 ppm has been observed in well waters (Buswell and Larson, 1939).

Atmospheric chemical constituents occurring either through wet deposition or dry deposition play significant role in determining water quality. The pH value would be towards acidic because of concentration of amino acid, NHO_3, and H_2SO_4, and oxides of nitrogen and sulphate, and due to the concentration of H^- ion. Moreover, H^+ ion in presence of Cl^- ion forms HC1. Formation of HC1 would be relatively small as Cl^- also reacts with Na^+. The solubility of HCl in clouds would increase acidity of water. Dry deposition of excessive CO_2 would neutralize the acidity upon its concentration, as it is alkaline in nature. Presence of Ca^{++} and Mg^{++} would increase the carbonate hardness of water, while Na^+ and K^+ ion because of alkaline nature would cause negative hardness. Excessive precipitation of trace elements with storm would increase contamination level of water (Table 16).

Table 16. Activities affect chemical parameters (Ghosh and Seth, 1994).

Parameters	Activity
Acidity	* dry and wet deposition of H_2SO_4 HNO_3, ammino acid, HC1, and SO_4^{2-} ppt.
Alkalinity	* dry deposition of CO_2,K^+ and Na^+
Ammonia - N	* wet deposition from canopy.
Calcium	* ionic precipitation with storm
CO_2	* dry deposition on soil
Chloride	* ionic precipitation or in the form of NaCl and HCl
Hardness	* precipitation of Ca^{2+} and Mg^{2+}
Magnesium	* ionic precipitation with storm
NO_3^--N	* wet deposition in soluble form of HNO_3,
	* dry deposition of HNO_3
Potassium	* wet deposition from canopy
Sodium	* ionic precipitation with storm or as compound of NaCl
Sulfate	* dry deposition or wet deposition of H_2SO_4

Thus we see that water pollution is caused by air pollution, but the system is much more complex than envisaged.

(b) Weathering of rocks, minerals and soils

All natural waters containing soluble inorganic ions are mainly from the weathering of soil and rock minerals. The weathering products of rock minerals are released and transported by water. Hence the nature and concentration of an ion in water depends upon the nature of rock mineral, its solubility and weatherability in fresh water or carbonated water (due to dissolution of atmospheric Carbon dioxide in rain water) climate. and local topography.

The main rock minerals forming the earth's surface are oxides of Silicon (quartz), Aluminium (gibbsite), Iron (huntite, limonite, magnetite). Titanium (rutite), and of sulphides (chlorite, pyrite), sulphates (gypsum, baryte), carbonates (catate, magnesite, dolomite), hydroxide (brucite), phosphates (apatites) and two-layer (kaolinite), three-layer (montmorillonite, illite, verculite) and interstratified clays.

These minerals on weathering release simpler radicals or ions such as Na^+, K^+, Ca^{+2}, Mg^{++}, HCO_3^-, SO_4^-, $Cl^-SiO_3^-$, $SiO_3(OH)^-_2$, $A1(OH)^-_4$, $H_2PO^-_4$, HPO_4^{-2}, $H_2P_3O_7^{-2}$, $FeHPO^-_4$, $C_3H_3PO^+_4$ etc. Among these Ca^{+2}, HCO^-_3 and H^+ and perhaps Mg^{+2} also are mainly controlled by the dissolution of carbonates rocks being easily weatherable and the source of Na, K and H_4SiO_4 and also Ca and Mg to some extent, is from the weathering of silicate minerals which are in contact with the ground water and the streams (Bricker and Garrels, 1967).

Systematic work on the solubility of minerals indicates the following decreasing order of weatherability: Gypsum > Calcite > Ca-feldspar > K-feldspar > Na-feldspar> Ca-and Na-montmorillonite > Quartz > K-mica > Gibbsite > Kaolinite > Hematite (Sturam and Morgan, 1970).

In natural waters, solubility of minerals is influenced by pH particularly of iron and manganese hydroxides which decreases and that of Aluminium hydroxide which increases with the increase of pH.

In highly polluted waters, containing organic materials, these released ions are likely to interact again forming organo-metallic complexes depending upon the nature of the environment. Several prossesses such as ion-exchange, fixation and release, chelation, precipitation and dissolution, oxidation and reduction, coagulation and aggregation are operating simultaneously which together make polluted waters a more complex system.

The complexity of the problem can be visualized in case of iron, manganese and phosphorus which are highly influenced by pH and Oxygen concentration. In the surface layers of natural waters. Oxygen concentration is high, and as soon as Fe^{+2} and Mn^{+2} (ions) are introduced into these layers, these are oxidised to insoluble form. The colloidal stability is enhanced by complex forming organic and inorganic bases (Hayes and Phillips, 1959; and Phillips, 1964).

(c) Decomposition of plant and animal material

Large quantities of soluble salts, and particularly nutrients are released from animal and plant residues or from added fertilizers. Decomposed organic matter and animal wastes are transported by water. It has been estimated that about 3905 million tons of soluble material is carried away annually from the continents by the flowing water (livingston, 1964). In India alone, about 5.37 million metric tons of nutrients (NPK) are annually washed away into the sea (Vora, 1975).

Nitrogen being a constituent of organic matter, on bacterial decomposition, is released in different forms such as NH^{-}_{4} , NO^{-}_{2}, NO^{-}_{3} , and N_2 according to the Oxygen status of the environment. Some nitrogen as nitrate, is also transported from the agricultural lands, and if it is allowed to accumulate in a water body its concentration increases.

Phosphorus is found in natural waters as orthophosphates, pyrophosphates, trophosphates, trometaphosphate, organic orthophosphate, phosphoamide, phosphoprotein, organic condensed phosphate and phosphorus containing pesticides. Stumm and Morgan (1970) discussed various physico-chemical processes as solubility, hydrolysis, phosphate equilibria at different pH, complex formation with metallic ions and sorption in natural waters.

When phosphate concentration in water exceeds 0.1 mg/l, as organic phosphorus, it proves harmful to algal growth while the critical limit for inorganic phosphorus is still 0.005 mg/l. It has been estimated that about 100 mg of algal biomass is synthesized by one mg of organic phosphorus (Lee and Hoadley, 1967).

The photosynthesis and respiration characteristics of the organisms appear to control sulphur status of water in reduced conditions, where SO_4 is reduced to H_2S by bacteria *Desulfovibrio* and by redox reaction it is deposited as elemental sulphur.

Thus, organisms play a vital role in influencing the water pollution through chemical changes in Nitrogen, Phosphorus and Sulphur compounds.

(d) Domestic and industrial effluents

The quality of water is deteriorated on the addition of used water having chemical constituents of undesirable concentration. In cities, contaminated water after bath, kitchen wash, washings etc. and a large volume of raw sewage discharged into the main stream pollutes water to a great extent. Similarly, discharge of municipal wastes, industrial effluents and wastes by-products aggravate the water quality (Mohan Rao, 1971; CPHERI, 1972). Chemical and biological characteristics of some common effluents show them to be quite high in their organic matter content (Table 17).

The extent of pollution depends upon the volume of effluents discharged per day and their characteristics. It has been estimated that various chemical industries need water upto 1100 m^3/tonne by pulp and paper industry, 45 m^3 both by steel and laundering, and 4-10 m^3/tonne by vegetable canning, sugar refining, coal mining, milk processing and baking industries are

Table 17. Characteristics of Selected Industrial Effluent and sewage of India

Industry	pH	Suspended solids (SS) (mg/1)	BOD (20^0C) (mg/1)	COD (mg/1)	Miscellaneous constituentis	Pollutional aspects
1. Sugar Factory	7.8-8.0	1500-1832	650-820	60-98	–	Large volume, high pH, SS, colour and toxicity.
2. Straw board	7.5-12.9	3000	2000	5000	–	High pH, SS and BOD
3. Tannery	9.5	3200	7000	–	Chromium 15-20 mg/1	High BOD, SS, Cr and colour
4. Cotton textiles	8.0-11.0	30-50	200-600	–	Detergents, dyes, chromium^3 mg/1	Alkali, BOD, dyes and varying chemical quality
5. Distillery	4.3	4000	29000	65000	Oil and grease	High BOD. SS, grease and ready putreseibility and low pH
6. Steel mill coke ovens	–	–	630	–	Grease, oil and tar NH_3/N 1000 mg/1, phenols 1300 mg/1 ONS	Highly toxic due to phenols Cyanides and ammonia
7. Steel (finished)	–	310	280	–	Phenol 97.5 mg/1, NH_3/N 440 mg/1	—do-
8. Refinery	–	–	200	–	180 mg/1 and sulphides, oil 30 mg/1 phenol 30mg/1.	Mineral oil and phenols
9. Fertilizer, Ammonia and Urea	8.0	3700	30	330	NH_5/N 510 mg/1 arsenic	Toxic due to free ammonia and promotes eutroptiication .
10. Dairy	8.0	690	816	1340	Oil and grease	High BOD, SS, grease and ready putreseibility
11. Sewage	7-8	200	350	500	Ammonia,Albuminoid	

Course Manual on Industrial Wastes Treatment NEERI. 1975

Table18. Characteristics of effluents obtained by different methods from the National Newsprint and Paper Mills Ltd., Nepanagar, M.P.

Table indicating the characteristics of effluents coming out from various sections in the Mill

Characteristics	*Mechanical pulp effluent	Chemical pulp mill effluent	Cold caustic soda pulpmill effluent	Paper machine effluent	Composite effluent	**Grade III
1. Colour	Light brown	Dark brown	Dark brown	Colourless	Light brown	Dark brown to black and muddy
2. pH	5.5 to 7.0	8.5 to 9.5	8.5 to 9.5	5.5 to 6.5	6.5 to 8.8	10.3 to 12
3. Total solids mg/l	1500-2000	900-1000	1500-2000	600-800	1500	800-2000
4. (a) Suspended solids	600-800	150-200	300-600	150-200	300	500-1000
(b) Dissolved solids	900-1200	750-800	1200-1400	450-600	1200	—
5. BOD mg/l	600-800	60-100	700-900	50-70	220	300-500
6. COD mg/l	1600-2000	300-400	1600-2000	150-200	700	700-1200
7. Total nitrogen (mg/l)	12	Nil	12	nill	5.77	—
8. Total phosphorus (mg/l)	3-5	Nil	3-5	nill	1.41	—
9. Sodium content (mg/l)	20-30	400-600	300-500	20-30	60-80	20-30 meq/1
10. Calcium content (mg/l) as $CaCO_3$	100-200	400-500	200-300	100-200	250-400	5-8 meq/1
11. Volumes of different effluents cu. m/day	10,000	36,550	12,500	11,500	70,500	

* Effluent 2-6 from NEPA Mills, Nepanagar provided by Das Gupta (19.76)

** Effluent 7 from Orient Paper Mill, Amlai provided by Late D. M. Prasad, NEERI, Nagpur

needed (Tebbutt, 1973). About 140 m / tonne water is needed for every 1000 metre of cotton cloth. This reflects that except some chemical industries, pulp and paper industry discharges maximum amount of polluted water. Consolidated data on the chemical nature of the effluents from Newsprints and Paper mill, Nepanagar, Orient paper mill, Anlai, reflect the quality and amount of water released per day (Table 18).

Central Pollution Control Board (CPCB, 1989) has indicated that a total generation of 12,145 million litres of waste-water per day out of which only 2633 MLD (22 %) were collected through sewerage system and rest were directly discharged on land or water bodies Without treatment (Table 19). From the total collection of 2633 MLD, only 859 MLD got primary treatment alone and 1626 MLD got both primary and secondary treatment. In the absence of tertiary treatment the released wastewater was capable of causing eutrophication of water bodies.

Table 19. Statewise Position of Wastewater Generation Collection and treatment in Class I Cities

State/ Union Territory	Total No. of cities	Wastewater (MLD)		Treatment (MLD)	
		Generated	Collected	Primary only	Primary and secondary
1	2	3	4	5	6
Andhra Pradesh	17	781	163.40	–	172
Bihar	11	445	–	–	43
Gujarat	11	815	220.5	86	463
Haryana	9	130	38.53	–	–
Jammu & Kashmir	2	209	–	–	–
Karnataka	14	655	462	327	10
Kerala	6	223	24.08	–	5
Maharashtra	29	2838	190	293	–
Manipur	1	207	–	–	–
Madhya Pradcsh	14	552	168	86	–
Orissa	6	207	22	–	–
Punjab	7	233	158	–	–

1	2	3	4	5	6
Rajasthan	11	318	–	–	–
Tamil Nadu	19	487	53	–	–
Tripura	1	13	–	–	–
Uttar Pardesh	29	1449	255	–	160
West Bengal	21	1107	27	–	20
Chandigarh	1	144	106	68	–
Delhi	1	1480	745	–	745
Pondicherry	1	38	–	–	–
Total	212	12145	2633	859	1626

Source - CPCB 1989.

In India, some 5000 large and medium industries were responsible for pollution of our waterbodies. But their contribution to the total pollution load was less than 10 percent. The rest was of domestic origin. For example, the industries of Mumbai accounted for only 13 percent of the total waste disposal into waterbodies and in Kolkata 11 percent of wastes were of industrial origin. In Delhi the river Yamuna took in 200 million litres of untreated human wastes everyday against 20 million litres of industrial wastewater. However, industrial effluents pose a greater threat due to the presence of toxic chemicals and non-biodegradable pollutants.

Waste water from municipalities are increasing due to expanding population and greater per capita water usage. Waste load from municipalities are expected to increase as the urban areas outgrow their treatment facilities. Only a fraction of the population is served with adequate sewers and waste water treatment facilities.

The growth rate of industrial waste water discharge is greater than that for municipal discharges. Quality characteristics of industrial wastes vary considerably depending upon the type of industry.

Agricultural wastes include irrigation return flows as well as runoff from fields. These waters exhibit salinities that are several times greater than unused irrigation water; also hardness, total dissolved solids, and turbidity are at increased levels. Irrigation return flows may also exhibit increase in nitrogen, phosphorus, pesticides contents, and micro-organisms that are potentially pathogenic to animals and human beings.

Soil erosion is one of the major water pollutants in terms of quality. Sediments from soil erosion is greatest in streams whose catchment area is denuded of forest cover.

Accidental spillage of oil and other hazardous substances into waterbodies is of fairly recent interest. There have been many instances of oil tanker collusions and offshore oil leak with devastating and extensive damage to the aquous environment.

Acid mine drainage deteriorate water quality of receiving waters. Problems are generated when water and air react with sulphur-bearing minerals in mines or refuse piles to form sulphuric acid. Acid mine drainage also contains iron, copper, leaf, zinc and other metals that may be toxic to aquatic life.

Watercraft discharge sanitary wastes, oils, litters, ballast and bilge wastes. Although the total quantity of waste discharged from watercrafts are small relative to other pollutant sources, they are important since most of the discharges are in high-use shoreline and harbour areas.

CLASSIFICATION OP WATER POLLUTANTS

Several attempts have been made to group water pollutants into classes or categories. Chaphekar and Mhatre (1981) and Das (1987) classified pollutants according to their mode of occurrence into physical, chemical and biological pollutants (Table 20). They have also discussed the nature of these classes of pollutants and have given their examples.

Table 20. Classification of water pollutants.

Occurrence	Nature	Examples
Physical	Temperature	Waste neat from industry.
	Turbidity	
	Colour	Dyes and pigments.
	Suspended & floating matter	Silt, sand, metal pieces, rubber, wood chpis, paper, foam, scum, carcasses, sewage.
Chemical	Inorganic	Nitrates, phosphates, chlorides, fluorides, salts etc.
	Organic	Detergents, tar, plastic, pesticides.
Biological	Pathogenic	Bacteria, virus, nematodes, worms, protozoans.
	Nuisance organisms	Slime, mollusca, algae, Ascellus, nematodes.

PHYSICAL

Physical characteristics which render water unacceptable to physical sences are included in this group. Some of the common and important characteristics are:

(a) Temperature

The discharge of unutilised heat into the cooling water source causes stratification, potentiating processes of uptake and degradation of organic matter. Heated effluents from power stations are discharged at 8-10°C higher than the coolant intake water. As a result, close to the outfall, there exists an intermixing zone with water temperatures slightly higher than ambient waters. The intermixing zone is followed by a diffuse zone with marginal increase of temperature in the waters. The temperature variations in the small area near the outfall are less than the natural fluctuations of temperature in the total water body around the year.

The heat exposes a paradox, though some heat is good some of the time, too much heat is always bad. A major problem caused by excessive temperature is oxygen depletion because of faster degradation of organic matter. At 5°C the maximum solubility is 12.8 mg/l of oxygen. The same at 40°C is reduced to 6.6 mg/1 (Table 21). Two factors associated with rising temperature are: decrease in available oxygen, and increase in metabolic and degradative rates. These two factors combine to render the aquatic environment less compatible to fish life at high water temperature.

If temperature rises rapidly, fishes die quickly depending upon the temperature and acclimation temperature. The sensitivity of fish to temperature can be obtained by substracting the accilation temperature from the lethal values (Table 22). The thermal death may be due to the action of heat on nervous system, coagulation of cell protoplasm or the inactivation of enzymes. They decrease as the acclimation temperature rises and indicate a level beyond which fish cannot be acclimated.

Table 21. Effect of temperature on water.

Temperature (°C)	Density (gm/cm^3)	Abs. viscosity (centipoise)	Pressure (mm Hg)	Dissolved oxygen at saturation (mg/1)
0	0.99987	1.7921	4.58	1.46
4	1.00000			
5	0.99999	1.5188	6.51	12.8
10	0.99973	1.3077	9.21	91.3
15	0.99913	1.1404	12.8	10.2
20	0.99823	1.0050	17.5	9.2
25	0.99707	0.8937	23.8	8.4
30	0.99567	0.8007	31.8	1.6
35	0.99406	0.7225	42.2	7.1
40	0.99224	0.6560	55.3	6.6

Table 22. Acclimation temperature and thermal death point for various fish species.

Fish	Acclimation temperature °C	Thermal death-point C°
Bluegill	15	30.7
Brook stickleback	25-26	30.6
Brown trout	26	26
Brown trout (fry)	56	22.5
Brown trout (fry)	20	23
Carp	20	31.34
Catish or brown bullhead	15	31.8
Chinook salmon (fry)	15	25
Chinook salmon (fry)	20	25.1
Chim salmon (fry)	15	23.1
Chim salmon (fry)	20	23.7
Common shiner	15	30.3
Common sucker	15	29.3
Creek chub	15	29.3
Pathead minnow	10	28.2
Pathead minnow	30	33.2
Golden shiner	15	30.5
Goldfish	10	30.8
Goldfish	20	34.8
Goldfish	30	38,6
Guppy	30	34
Largemouth bass	20	32.4
Largemouth bass	30	36.4
Perch	—	23.25
Pink lalmon (fry)	5	21.3
Pink salmon (fry)	10	22.5
Pink salmon (fry)	20	2.39
Pumpkinseed	25-26	34.5
Rainbow trout	—	28
Rainbow trout (Kamloopsvar)	11	24
Roach	20	29.5
Roach	25	30.5
Roach	30	31.5
Sockeye salmon (fry)	5	29.8

Major thermal discharges originate from thermal power producing projects. For instance, an average 335 MW conventional coal-based power station requires 950 m^3 water per minute prducing a temperature rise of 12.8°C between intake and 3 outlet in winter; in other seasons it uses upto 1900 m^3 water per minute increasing the water temperature 6.4°C above ambient (Mihursky, 1969). Nuclear power stations require more cooling water per MW than fossil fuel power stations (Parker, 1965) and produce about 50 percent more waste heat per unit of electricity. However, in many countries legislative regulations limit the effect of thermal addition to 3-5 °C (Belter, 1975).

The heat input of a fresh water body is one of the driving forces for the assimilatory capacity. Its impact is modified by the way in which it is introduced. In shallow waters, the seasonal characteristics may be severely disturbed by thermal discharges. In stratified waters heat addition may cause prolonged thermal stratification and delayed turnover in automn with its consequences for oxygen and nutrient dynamics of the water body. Since thermal discharges reduce the capacity of the water to keep oxygen in solution and at the same time often raise the biological oxygen demand (BOD). Both these factors decrease the assimilatory capacity of the aquatic ecosystem in general (Ross, 1977; Stangenberg, 1977).

Sensitivity of plants to temperature depends on life-cycle and species. Elevated temperature in automn may cause disturbance in dormancy in *Ceratophyllum demersum* and *Myriophyllum spicatum* (Vegis, 194t a,b: Weber and Nooden, 1976; Best, 1979). *Ruppia maritime* anthesis come to a complete standstill at 25°C (Setchell, 1924). *Potamogeton perfoliatus* increase in abundance indicating tolerance for elevated temperature of 35 °C (Anderson, 1969).

A few studies indicate changes in population structure in mixing zones of thermal effluents. Grace and Tilly (1976) found that the biomass of submerged plants increased at the warm stations (23.8 - 30.5°C) as compared to the cold stations (21.1 - 30.5°C) largely due to multiplication of *Myriophyllum spicatum.* At the hot stations (26.3 - 33.4°C) *Myriophyllum spicatum* was outcompeted by *Najas gradulupensis.* Haag and Gorham (1977) noted replacement of *Myriophyllum exalbescens* and *Chara globularis* by *Potamogeton pectinatus* and *Elodea canadensis* in the above such mentioned mixing zones.

Thermal pollution has been the subject of much debate. Extensive research has been conducted into both the good and the bad effects of discharging heated coolent water. Various species of fish respond differently to heat. It has been found that 28°C is about the maximum which would support growth of catfish, carp and other "tough" species. Lake trout, atlantic salmon and other salmonids requires water no warmer than 14.2°C for spawning and egg development. Pacific Gas and Electric Company diverted heated coolant water over the spawning beds and nursary grounds in Humboldt Bay, where it was found that this practice improved growth of both calms and oysters. Counterimg this, it has been found that pathogenic bacteria like warmer water too.

Moreover, rapid changes in temperature can create lethal shock effects, for example, temperature is either increased or decreased, greater than 0.29°C per hour, can create a lethal effect on fish. There are many cases on record where fish have been attracted to and become acclimatised to flow of heated coolent water below Thermal Electric Generating Stations. Later when the unit have been shut down for repairs and cold water has replaced the hot water flow, the shock has killed the fish in the discharge zone.

Migrating fish requires zones of passage, and if the flow of heated coolent water or other pollutants block these passages, the effect creates a barrier which is as effective as dams in blocking migration.

(b) Turbidity and water colour

Turbidity and water colour can be regarded as aesthetic pollutants. Aesthetics embraces all aspects of human enjoyment which depend on looking and thereby enjoying the appearance, odour and contact (feel) of water and its surrounding natural environment. Aquatic recreation embraces swimming, water skiing and all related water sports. Fishing is one of the most universal forms of recreation.

Turbidity and water colour are imparted to water due to certain inorganic and organic compounds being introduced into it either through domestic or industrial effluents. Turbidity and water colour are serious pollutants for several reasons. One of the most serious impact is that they block sunlight from reaching algar and other aquatic vegetation. The result is that photosynthesis is retarded. Turbidity levels govern the depth of the *euphotic zone.* This is the depth at which 99.7 percent of the solar radiation has been attenuated, absorbed or scattered. Most of the primary productivity of the oceanic biomass is concentrated in this zone, furnished by phytoplankton. Quantitatively these organisms are more important than the bottom algae and the rooted aquatic plants.

Turbidity can have a drastic effect on calms. *Mercenaria mercenaria* was subjected to turbidity tests, 3 gm of non-toxic sediment was added to 1 litre of water, in even this concentration, all test animals died. Turbidity often stem from masses of free swimming organisms. The Euglenophytes are well known to cause "red tide" conditions in the aquatic environments.

(c) Suspended and floating matter

Most of the suspended and floating matter include silt, sand, metal pieces, rubber, wood chips, paper, foam, scum, wood, carcasses and sewage, which stem from human concentration near the water bodies.

When our primitive ancestors lived in widely separated family units, human bodies and household wastes could be disposed of easily. Later when settlements developed, problems arose rapidly. The Chinese solution was to haul human body and wastes out of the town in carts, dry them in sun, and use them for fertilizers. Most of the pathogenic organisms died in this way, and although others were strong during collection and handling, they were soon killed by the sun. The European and later American solution was to convert human body into liquid form or fine suspensoids so that they could be drained off or flushed away in ditches and gutters allong the roads. The wastes were diverted into pipes underground and discharged into the streams.

Pollution is most often caused by the uncontrolled disposal of sewage. It is liquid from residential and industrial establishment of a town. It is made up of spent water from bathrooms, lavatory basins, kitchen sinks and the discharge from urinals of homes, offices and institutions; human, animal and bird feaces. The sewage is extremely foul on account of the human excreta (or night soil) and urine contained in it, which putrify and give out offensive smell. In sewage polluted water, the structure of algal communities undergoes several changes. Heterotrophic forms like *Euglena viridis, Nitzechia palea,* and *Oscillatora limosa* become important. These algae help in supplying the much needed oxygen to the large population of bacteria that breakdown organic matter. In the absence of oxygen anaerobiosis occurs in which fermentation of carbohydrates and sugars and putrification of proteins takes place. In such a breakdown of proteins. Hydrogen sulphide is produced, which bubbles out giving the smell of rotten eggs. The water body becomes a source of nuisance and fish and other animals are killed.

CHEMICAL POLLUTION

The changes brought about in the chemical characteristics through the domestic and industrial effluents causes chemical pollution. There can be innumerable types of chemicals which are dumped in the water bodies. Some of the important ones are:

(a) Nitrates

Ganapati (1960) pointed out that tropical waters, particularly the non-polluted ones are deficient in nitrogen. The concentration of nitrates was much higher (20 mg/1 to over 300 mg/l) in well waters of Canada (Robertson and Riddel, 1949). While it was 10 - 257 mg/l in Czechoslovakia (Schmidt and Knotek, 1970). Zafar (1964) observed that the dead organic matter gets converted into nitrogenous organic matter and finally to nitrate by bacterial activity in polluted waters.

(b) Phosphates

In natural waters extremely small quantities of phosphorus are often found and any amount in excess of 0.5 mg/l has been regarded as an indicator of pollution (Atkins and Harris, 1924). The total phosphorus in lake waters ia known to vary from an undetectable amount (1 mg/m^3) as in the surface waters of the Traunsee (Ruttner, 1937) to intense quantities (789 mg/m^3) in Owens lake, California. High phosphate contant might by attributed to drainage in the lakes. The unfiltered water of the catchment area of phosphate rocks and uncontrolled disposal of sewage and biodegradable synthetic detergents also add huge quantities of phosphate.

(c) Chlorides

In uncontaminated waters which are not flowing from saline sediments, the quantity of chlorides present is ordinarily very low (Hutchinson, 1957). High concentration of chlorides during summer months might be due to decomposition of organic matter as advocated by Adoni (1975).

(d) Fluorides

Surface waters generally contain less than 0.5 mg/l fluorides. However, when present in much greater concentration, it becomes a pollutant. Areas exist where the fluoride content of water ranges from 1.5 to 6 mg/l, for example, in the Kurnool district of Andhra Pradesh (Ramamohan Rao and Bhaskaran, 1964), upto 12 mg/l in New Mexico (Adler et al, 1970). Well waters of several villages of Nagaur and Churu districts of Rajasthan contain 1.1 to 48.0 mg/l fluoride (Purohit, 1917).

(e) Toxic elements

Many toxic elements to a large extent, are dispersed in the aquatic ecosystem through industrial effluents of mines discharge, metal processing, calico printing, organic wastes, refuge burning, transport and power generation. Some of the toxic elements are discussed here.

Cadmium levels range from less than 1 µg/l to 10 µg/l in natural waters. In addition to industrial discharge, metal or plastic pipes constitute a possible source of cadmium in water. Like other metals in solution, cadmium tends to be absorbed by suspended particles and in the bottom sediments. For this reason, even in polluted rivers, the cadmium levels in the water phase may be below the detactable limit. Irrigation water containing suspended solids contain very high cadmium concentration.

Lead in water mainly comes from lead processing industries, or due to the use of lead pipes. Natural and untreated water supplies contain about 0.01 to 0.03 mg/l of lead. Kopp and Kroner (1970) reported that only 2 percent of samples from rivers and lakes in United States had lead concentration in excess of 0.05 mg/l. Problems exist, however, in areas with soft, slightly acidic water, which may dissolve lead from the lead pipes, plastic pipes in which lead has been used as a stabilizer.

Mercury is one of the worst offenders as a dangerous pollutant among the heavy metals. Mercury is discharged in the inorganic form from mercury and other metal mines. Man-made sources of mercury include mining and refining processes, paper and pulp industry, acetylene-acetaldehyde synthesis, vinyl chloride synthesis, caustic soda industries using mercury cell, organo-mercuric fungicides and seed disinfactant, mercury electric appliances industries etc. The level of mercury in natural fresh water are generally below 0.1 μg/l, but in contaminated areas such as Minamata bay in Japan, mercury levels as high as 1 to 10 μg/l were observed (Kiyoura, 1963). The mud of Minamata bay was found to contain from 30 to 40 μg/kg of mercury (Kitamura, 1968). Mercury found in industrial effluents reach bottom sediments, where they are metabolically converted into Methyl mercury by the activity of anaerobic micro-organisms. Methyl mercury is highly persistent kind of pollutant that accumulates in the food chain. Mercury pollution has been reported from various places in the world, such as Brazil, Canada, Germany, Holland, Italy, Japan, Russia and United States. There is possibility of mercury pollution in Poland, China and India too.

Heavy metals produce toxic effects and endanger the life of aquatic fauna of which fishes are the most sensitive group. Copper when absorbed through gills, result in considerable tissue damage and becomes fatal to fish life (Doudoroff and Katz, 1953).

Cadmium enters a fish body through the absorption by general body surface. Zinc is taken up through the oral route by fishes.

Gautam and Lal(1988) studied the effect of Copper, Cadmium and Zinc to a larvicidal fish *(Gambusia affinis patrulis)*. It was observed that 72 hour and 96 hour LC_{50} values and their 95 percent confidence limits in mg/l for metal ions were 15(11-18) and 11(8-14) for Copper, 26.5(24.2-27.5) and 22.0 (21.0-23.5) for Cadmium, and 33.5 (31.3-34.0) and 27.5(26.0-28.0) for Zinc. These results indicate that Copper was most toxic and Zinc was least so amongst the test metal ions. The relative mortality of Copper was more than that of Cadmium and Zinc. Most of the fishes appeared physically normal until the time they died, but some became increasingly disposed towards convulsions to the point of being sent into a frenzy of vigorous swimming each time they were disturbed. The peroxysms generally terminated in immobilization of the fish in tetany on the tank bottom for several seconds or minutes or sometimes in the death of the fishes.

The toxicity of various salts of Zinc on a number of fish species have been demonstrated by McKee and Wolf (1963), Pickering and Henderson (1964), and Khangarot (1980). Gautam and Agarwal (1987) observed acute toxicity of Zinc at two different temperatures of 18.9-19.6°C and 32.0-34.2°C on *Cirrhinus mrigala*. The concentration of 70 mg/l of Zinc caused 100 percent mortality in 6 hours at higher temperature, while at the same concentration 100 percent mortality appeared in 36 hours at lower temperature (Table 23). Fishes exposed to Zinc ions showed abnormal behaviour as compared to control ones. Before death they passed through various stages such as: (i) come to the surface repeatedly, (ii) did not live in shoal, (iii) lost their equilibrium, (iv) swim with belly upwards and sank to the bottom, and (v) finally before death showed irregular opercular movement. It was also observed that in higher concentration, mucous started precipitating on the gills and general surface of the body.

Other toxic elements and their compounds are found in natural and polluted water, usually in very small concentration. They include Barium, Berryllium, Molybdenum, Tin, Uranium, Vanadium etc.

Table 23. Concentration of soluble Zinc (mg/l) and percent mortality in *Cirrhinus mrigala* at different time interval and temperatures.

Calculated zinc	Observed zinc		Mortality at different hours		
	Mean	Range	24	48	96
Temperature	18.9-19.6°C				
70	20.5	19.90-20.70	90	100	100
54	17.0	16.70-17.20	60	70	80
42	15.0	14.90-15.30	40	60	60
33	11.5	11.30-11.70	10	20	55
20	8.0	7.85- 8.40	0	0	15
Control	0.4	0.25 - 0.50	0	0	0
Temperature	32.0-34.2°C				
42	12.0	11.90-13.00	100	100	100
33	9.80	9.70- 9.90	80	100	100
20	3.10	3.00- 3.80	80	90	100
15	2.30	2.25- 2.35	40	70	80
10	1.57	1.53- 1.59	30	40	50
5	1.00	0.87- 1.10	0	0	10
Control		0.20 - 0.35	0	0	0

(f) Synthetic detergents

Sodium alkyl sulphate (AS) is a synthetic detergent commonly included in dish detergents, shampoo and commercial toothpastes. It is used as a surface active component causing foam. Residues of Alkyl benzene sulphate (ABS) are non-biodegradable Linear Alkyl sulphate (LAS) are biodegradable. The most effective and most widely used compound today is Sodium tripolyphosphate (STP). Enormous quantities of these synthetic detergents are added to water bodies, because of their increasing use in domestic and industrial processes.

Synthetic detergents are toxic water pollutants. Concentrations as low as 1 ppm are toxic to fish which are exposed for

24 hours or more. Most of the currently used detergents are biodegradable, that is, they can be consumed by organisms. When detergents decompose and are reduced to the basic inorganic chemical components, they may cause serious additional problems. Detergents contain much phosphorus, one of the most important fertilizers. When added in water, it stimulates algal and aquatic plant growth. Photosynthesis is generally accelerated during day light hours. These processes may stimulate additional heat. When light falls on them, the plants produce oxygen, frequently exceeding 100 percent saturation.

Unlike soap detergents are not easily decomposed by bacteria and are richer sources of phosphate that hasten the process of eutrophication of water with frequent algal blooms (Agarwal, 1982). Some of these blooms persisted for a week or more while others lasted for a year or more. These algal blooms impart colour, odour and unpleasant taste to the water and cause inconvenience in water supply, fish farming and make the use of water ineffective (Seenayya, 1980).

(g) Pesticides

Several groups of organochlorine compounds are of interest in water pollution. They include insecticides such as DDT, aldrin, endosulphan, chlorinated phenoxy acid, used as herbicides, fungicides such as hexachlorobenzene and penta-chlorophenol and polychlorinated biphenyl (PCB).

There are two sources of organochlorine pesticides in water, run-off from agricultural land and discharge of industrial water. Because of their low solubility in water and tendency to be absorbed on solid surfaces, only traces of these compounds are found in raw water and treated water. Residues reported in water samples (Table 24) are mainly of agricultural insecticides on particulate matter suspended in water. Much larger quantities may be found in mud and bottom sediments. Major river basin in United States (Edwards, 1970) and United Kingdom (Lowden et al, 1969) show much larger values than those of Thailand (Niyomkha and Wanghanchao, 1975).

Table 24. Mean concentration (nanogram/litre) of pesticides in river basins.

Pesticides	USA	UK	Thailand
P,P' DDT (ppb)	–	–	0.92
P, P' TDE (ppb)	–	–	2.00
P,P' DDE (ppb)	–	–	0.10
DDT	8.2 - 10.3	1.6 - 64.6	–
Lindane	2.8 - 28.0	18.7 - 38.6	–
Aldrin	0.2 -5.0	–	–
Deldrin	2.3 - 10.0	3.3 - 114.0	–
Heptachlor	0.1 - 6.3	–	–
Endrin	1.4 - 541.0	–	–

Pesticides cause widespread pollution of various fresh water bodies like rivers, lakes and estuaries. Edwards (1973) compiled data on pesticide residues in rivers (Table 25). Most of the surveys generally refers to organochlorine insecticides. DDT residues were most common and often accounted for the largest amounts found. This was followed by Lindane, Dieldrin, Endrin, Heptachlor and its epoxide. DDT, Dieldrin and to some extent Lindane were seen in relatively lesser number of samples. Apparently, there has been an increase in the residue levels which reached a peak, sometimes around 1966 and thereafter there has been a decline both in terms of residue levels and occurrences. This may be the result of gradual decline in the use of organochlorine insecticides and use of carbamates and organophosphates as substitutes. Quite significantly, the residues of these substitutes pesticides have not shown much of an upward trend in the environment (Edwards, 1977).

The Great Lakes in the United States have been reported to be heavily contaminated with DDT. In Utah lake 1 ppb or more of Aldrin, Lindane and Heptachlor were reported (Bradshaw et al, 1972). As rivers contain significant amounts of pesticides, it is expected that these chemicals will be carried to the sea and there might be an accumulation of these chemicals in the estuaries. However, contrary to such expectations, relatively low quantities of pesticides have been reported from such habitats. In Lauisiana

Table 25. Residue of organochlorine pesticides in water (Agarwal,H.C. 1983).

Type of water body	Concentration of residues ppm x 10^6 (ng/1)									Reference
	Aldrin	Lindane	Chlor-dane	DDT	DDE	Dieldrin	Endrin	Heptachlor & its epoxide	Eado sulfan	
Rivers (UK)	–	53.6	–	8.67	–	2840	–	–	–	Croll, 1969
Rivers (West Germany)	–	138.2	–	18.9	–	3.2	–	39.4	126.7	Herzel, 1970
Mississipi river delta (USA)	5.0	28.0	–	112.0	–	10.0	541.0	2.0	–	Anon, U.S.D.A., A.R.S.. 1966.
South Skunk River (USA)	–	–	–	–	2 to 1820	2 to 76	–	–	–	Richard et al., 1975
Rivers (USA)	–	2.2	0.1	8.3	–	5.9	3.6	0.1	–	Green et al., 1966.
Streams (Western USA)	0.6	0.5	–	9.3	–	1.1	0.3	1.4	–	Manigold & Schulze, 1969.
Water areas (Ontario Canada)	–	–	–	64.0	–	7.1	–	–	19.6	Miles and Harris, 1971

estuary less than 1 ppb of Dieldrin and Endrin were found in water (Rowe et al, 1971). One of the reasons for low residue levels may be that the pesticides were adsorbed onto the particulate matter which eventually settle at the bottom.

Persistent organochlorine pesticides like DDT and lindane have been extensively used in India for a long time. Observations (Table 26) show DDT residues in various water bodies in and around Delhi. In the river Jamuna, levels of DDT residues were comparatively high downstream Wazirabad, which was reduced to the normal levels at Okhla. The total DDT residues included pp'DDE, pp'DDD and op'DDT in addition to pp'DDT. In Hindon river the residue levels were reasonably high. DDT was also detected in the water from Surajkund and Damdama lake. DDT was detected from one of the small ponds on the Delhi University Campus, though the concentration was small as compared to other places (Pilial et al, 1977).

Table 26. Mean levels of DDT in water from areas in and around Delhi from May 1976 to April 1978 (Agarwal,H.C. 1993).

Type of water body	Site	DDT. residue (in ppb)	Reference
Rivers			
Jamuna	Wazirabad upstream	0.249	Mittal et al. unpublished work
	Wazirabad downstream	0.558	"
	Okhla upstream	0.36	"
	Okhla downstream	0.294	"
Hindon		0.296	"
Lakes	Suraj Kund	0.195	"
	Damdama	0.377	"
Pond	Delhi University campus	0.064	Piliai et al., 1977.

DDT is insoluble in water, not leached easily and become bound to the organic matter or fractions of the bottom sediments and settle at the bottom. It has been reported that such pesticides tend to disappear very rapidly from the water and finally end up in a situation where much larger residues are present in the mud than in water above it (Rowe et al, 1971; Miles and Harris, 1971). Observations of values of DDT in samples of bottom sediments of river Jamuna (Table 27) indicate that the concentration of DDT and its metabolites was lowest at Wazirabad upstream, and increased gradually upto Okhla downstream, where they attained maximum levels.

Table 27. DDT and its metabolites in the bottom sediments of river Jamuna at Delhi during 1976 to 1978 (in ppm).

Site	p,p' -DDE	o,p' -DDT	p,p' -DDD	p,p' -DDT	Total DDT
Wazirabad upstream	0.017	0.002	0.031	0.058	0.108
Wazirabad downstream	0.023	0.033	0.065	0.133	0.254
Okhla upstream	0.031	0.021	0.059	0.094	0.205
Okhla downstream	0.037	0.062	0.077	0.177	0.353

Mittal et al., (unpublished work)

Bioaccumulation of pesticides present in water occurs in the tissues of plants and animals. The extent of this accumulation gives an indication of the pesticide transport in the food chain and also the potential hazard in the environment. Protozoans like *Blepharisma intermedium* accumulated DDT by about 3000 fold in 12 hours ver rapidly (Table 28), when exposed to water containing 0.1 ppb of pp'DDT were able to concentrate it one hundred fifty thousand times in 48 hours.

Table 28. DDT and its metabolites in *Blepharisma* intermedium exposed to 1 ppm of pp'DDT (in ppm).

Time of exposure	p, p' -DDE	o, p' —DDT	p, p' -DOT	Total DDT
1 hr	15.27	17.34	53.55	86.16
2 hr	8.89	90.42	544.29	643.6
3 hr	12.98	279.58	1006.5	1299.1
6 hr	10.96	51.94	1445.1	1508.0
12 hr	28.71	39.72	2991.7	2985.8
24 hr	36.24	137.96	2675.9	2850.1
48 hr	17.47	51.72	2835.1	2904.3
72 hr	19.34	61.25	3540.4	3620.9
5 days	22.84	44.89'	3293.0	3360.0
10 days	67.27	45.33	5887.3	5999.9

Ruplal et al., (unpublished work).

Observations (table 34) indicates residue levels of DDT in the water of a pond on the Delhi University Campus and the most ito fish *(Gambusa affinis)*. DDT was concentrated about 25,000 fold in the fish when exposed to very low concentration of DDT in water. The LC_{50} calues of *G. affinis* to DDT were found to be 60 parts per billion.

Table 29. Residue levels of pp' DDT and its metabolites in the pond water and *Gambusa affinis* (in ppm)

DDE	DDD	DDT	Total DDT	Concentration factor
		Pond water		
0.0000215	0.000024	0.0000185	0.000064	–
		G. affinis		
1.3826	0.1475	0.0887	1.6138	25,293
±0.0117	±0.0198	±0.0122	±0.1625	

From Pillai et al.,(1977)

(h) Tar pollution

Large scale oil spills alone do not spell doom for marine life. Crude oil leaking from tankers or flushed out of them poses a continuing threat. After more volatile parts of the oil evaporates, the remainder concentrates into tar. Blocks of such tar floating of Florida and in the Carribbian are being mistaken for food by Hawksbill turtles, an endangered species. As a result dead and dry tar laden Hawksbills are washing up on shores of the region. Oil pollution can have deadly effect on both freshwater and marine organisms. It removes vast amount of dissolved oxygen.

(i) Plastic peril

Plastic may fee as great a source of danger to marine mammals as oil spills. The world's merchant ships which are the prime contributors to the growing plastic pollution, dump atleast 6.6 million tons of trash over - board every year. Some 6,39,000 plastic containers and bags are tossed into the ocean every day. Commercial fisherman are also major offenders. Estimates of plastic fishing gears lost or discarded at the sea every year is as high as 1,50,000 tons. Boaters and beachers add to the marine trash with picnic utensils, sandwich bags and paper cups. Surveys have shown that plastic accounts for over one half of the man-made products on the ocean surface. Plastic is taking a heavy toll of marine life. 50,000 northern fur seals in the Pribilof island die each year after becoming entangled in nets. Smaller plastic items are frequently mistaken for prey by turtles and birds, leading to fatal consequances. Plastic either obstructed the digestive tract or caused ulcers

BIOLOGICAL POLLUTION

The biological organims cause water pollution. Some of them are pathogenic organisms while others are nuisance organisms.

(a) Pathogenic organisms

The contamination of water by pathogenic bacteria, viruses and parasites (Table 24), toxic and other kinds of algae and aquatic vegetation cause additional undesirable results. These organisms are found in the water source itself or during the conveyance water from the source to the consumer.

(b) Nuisance organisms

Five kinds of nuisance organiums (Table 37) have their greatest impact on aesthetic value and must be kept out of water. These are pollutants which :

(i) Settle and form objectionable deposits,

(ii) Form mats or aggregations of slime, scum, oil, foam etc.

(iii) Produce objectionable odours, taste, colour or turbidity,

(iv) Are toxic or which produce undesirable physiological response in human , fish, other animals and plants.

(v) Remove oxygen excessively, stimulating growth of sludge worms and other creatures.

Table 29. Nuisance organisms normally found in drinking water (WHO, 1977)

Organisms	Effects
Biological slimes	Chocking of treatment plants and distribution system. Support for methane utilising bacteria. Risk of rendering water unacceptable.
Mullusca	Chocking of water mains.
Plumetella, algae	Interference with filteration.
Asellus	Risk of rendering water unacceptable.
Nematodes	Possible concentration of pathogens.

(c) Microbial organisms

While high counts of faecal coliforms in water usually indicate heavy and recent pollution, their absence does not guarantee that the water is free from faecal contamination, since coliforms die rapidly in water. Of greater public health concern is the level of contamination with *Enterococcus faecalis,* because the latter does not multiply readily in water, they die less rapidly than faecal coliforms, and tend to persist even after clorination (WHO, 1984). As a result, the presence of *Enterococcus faecalis* in water sources is an indication that the water sources are being polluted by faeces at a distant location (Kenner, 1978). Ehiri et al

(2001) observed that inadequate supply of chemicals owing to lack of financial resources and poor maintenance of equipments are likely to be important determinants of the quality of tap water in many developing countries. It should also be noted that in many of these countries the primary challenge is the lack of water, rather than its quality (Cairncross, 1990). The outdoor defecation by childern and adults can contaminate water sources and may explain the high levels of pathogens in nearby streams.

Drinking water samples from Nigeria, 84 percent of the tap water samples did not meet WHO standards, and 33 percent of the samples from borewells also failed to meet the respective recommended standards. All stream samples showed contamination with faecal coliforms and *Enterococcus faecalis.* (Table 30). According to WHO guidelines, chlorinated water should contain no coliforms (WHO, 1984).

Table 30. Microbial contaminants of domestic water.

Organism	Pipe-borne n=19		Borehole n=6		Stream n=9		Spring n=8	
	Mean (range)	SD[b]	Mean (range)	SD	Mean (range)	SD	Mean (range)	SD
Faecal coliforms	8.74 (0-30)	8.37 (0-10)	2.50 (2-35)	4.18 (0-10)	12.33	10.78	3.13	4.18
Enterococcus faecalis	31.58 (0-80)	26.51	7.17 (0-20)	7.22	140.00 (12-250)	106.30	6.25 (10-20)	6.94
% of sources contaminated	84	68	33	83	100	100	25	63

[a] Measured as the number of microorganisms per 100 ml sample.
[b] SD = standard deviation.

Toxic and nuisance biological organisms cause additional undesirable results. Some blue-green algae are extremely toxic to livestock. Those of the genus *Aphanizomenon, Anabaena, and Anacystis* have been identified for numerous animal deaths. The relationship between polluted water animal death is well established. One of the best examples of water borne disease organisms which affect livestock in 'bacillary hemoglobinuria'. 'Anthrax' another deadly killer of livestock is caused by the organism *Bacillus anthracis.* Water pollution has shocking effect on wildlife. Most of the species which have become extinct within the recorded history have done so because of pollution of some sort. Wildlife requires water of a quality adequate to maintain its health. A healthy animal is one which can survive a normal span, display normal habits and migration and can reproduce successfully.

(j) Industrial pollution

Industry creates more pollution than does any other segment of the society. It has been observed that the pollution load contributed by a small scale industry may be equal to that contributed by the sewage from large city. Water pollution is most often caused by uncontrolled rejection of wastes (Table 31).

Table 31. Volume of waste water discharged from selected industries in India (upto the year 1980).

Industry	Unit	Volume of waste water per unit products manufactured (gallons/unit)
Pulp and paper	Ton of paper	50000 - 100000
Straw Board	Ton of board	20000
Tannery	100 kg of hide	750
Citten textiles	1000 meter of cloth	33000- 99000
Distillary	1000 litres of molasses	1320
Dairy	1000 litre of milk	440-1760
Steel mill ckeoven	ton of steel	125-155
Refinary	Ton oil produced	350-400
Ammonia fertilizer	Ton of ammonia	2000
Urea fertilizer	Ton of urea	1500

The nature and quantity of wastes differ from industry to industry depending upon the nature of raw material, process used, products manufactared and by products recieved.

Among all refining, mixing, blending and extraction, industrial manufacturing operations produce some wastes. This must be removed, and transported to some place for subsequent disposal. During this process accidental spills, leakage and so forth, which may occur inspite of out best efforts lead to grave consequences, perfection seems never to be reached, so atleast some waste are bound to be produced as long as industrial production continues.

CONSEQUENCES OF WATER POLLUTION

Water pollution leads to grave consequences as shown in Figure 14. These include the changes in the water quality itself, which make it unfit for domestic, industrial, agricultural and recreational use. All forms of aquatic life suffer from some sort of water pollution. However fresh water, marine and estuarine organisms vary widely in their ability to tolerate different kinds of pollution. Factors which may be safe for some may be deadly to others.

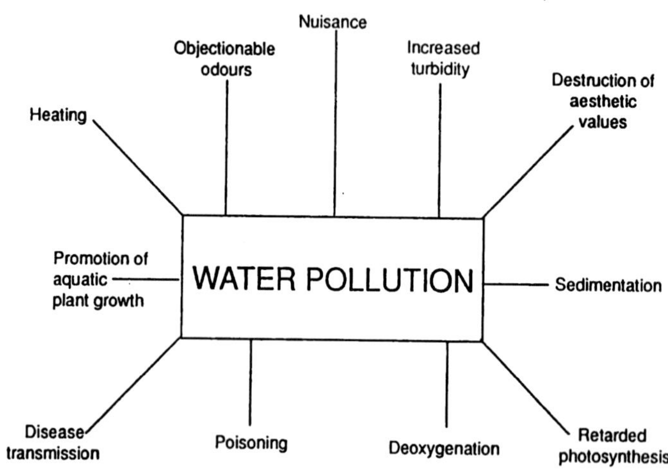

Figure 14. Various consequences of water pollution.

WATER POLLUTION SURVEYS IN INDIA

Despite the growth of large cities and towns, most people in India still live in rural communities. The situation in the larger cities such as Mumbai, Kolkata and Delhi is better. In these cities piped supplies of filtered and sterlized water are generally available, sewage treatment works exist, although these are often seriously over-loaded because of the rapid increase in urban population over the last two decades. Even in the bigger cities, control over discharge into rivers are so few and ineffective that almost no pollution control is possible. We wake up only when dead birds and floating fish in rivers throw up a warning that pollution has far exceeded the permissible limits. This has happened in Delhi at Okhla.

According to the survey carried out by several workers on selected stretches of some of the important rivers, it has been observed that the water of most of these rivers is polluted. A few of them have been discussed here.

The river Ganges is one of the first ten mighty rivers of the world. It originates from Gangotri in the Himalayan mountains and after flowing for about 2525 Km falls into the Bay of Bengal near Calcutta. The Ganga drains eight states - Himachal Pradesh, Punjab, Haryana, Utter Pradesh, Rajasthan, Madhya Pradesh, Bihar, West Bengal and the Union territory of Delhi. It passes along 29 class I cities (population between 50,000 - 1,00,000) and about 48 towns having less than 50,000 population. The Ganga waters are embodiment of more familiar recreational uses such as bathing, boating and fishing; non-recreational uses such as hydroelectric power project, water supply (including irrigation), transportation, industrial use and to dump their wastes into the same. The river serves as the source of water supply for these towns. Several major pilgrim centres have existed on its banks for centuries and millions of people come to have a holy dip in the river during religious festivals, especially Khumbhs of Haridwar and Allahabad. The Ganga however, is being grossly polluted especially near these towns (Figure 7). In the towns where there is no severage system or the coverage is only partial, the water flows through open drains and finally finds its way into the river.

About 9028 M^3/day of antibiotics waste from antibiotic plant, Virbhadra, Rishkesh is discharged in this river after treatment. Owing to large quantity of water available for dilution, this waste has meagre effect on water quality of this river. The dissolved oxygen value 6.0 to 6.5 mg/l and BOD value of 1.5 to 3.0 mg/1 does substantiate the above view, Kanpur has about 45 tanneries, 10 textile and several other industrial units directly discharging their untreated effluents in this river. The volume and BOD load of these wastes are approximately 37.15 million gallons and 61630 kg/day respectively. This has resulted in a significant degradation of water quality in the river Ganges with BOD values as high as 230 mg/1 and bacterial counts of the order of 400 to 2400 ppm/100 ml. This situation continues even down stream of Kanpur upto Jajamou where the bypass channel from the main sewage pump station meet the river. Colossal amounts of both biological and chemical filth are contributed to the river in Varanasi. Six million devoteesbath every year in the 7 km curve and one can well imagine the amount of the waste added to the waters. Besides this, as much as 60 million litres of untreated sewage is dumped by 6 major and 61 small drains in the city. The ashes of about 40,000 human bodies and about 10,000 half burnt bodies are also thrown into the river every year. In addition about 60,000 carcasses of cows, buffaloes and dogs are dumped annually into Ganga waters at Varanasi. In Bihar, near the Mukhmah and Barauni, the wastes from tennary, oil refinary and fertilizers are discharged, Thakur and Deshpande (1976) indicated that in 1968, heavy discharges of oil bearing wastes from the oil refinary of Barauni caused noticeable pollution in this river up to Monghyr. At Calcutta Hooghly river formed by three main distributaries of Ganga river viz., Bhagirathi, Jaiaugi and Churmi receives 252 million gallons of industrial wastes along with sewage per day which in terms of BOD, total solids, suspended solids and dissolved solids are 107.8, 2345.4, 1073.9 and 1169.5 metric tons/ day respectively. Due to these substances the condition of water below the outfall is found detriorating particularly in summer season.

Physico-chemical characteristics of five major draind at Patna which empty into river Ganga (Table 32) show that samples from Krishnaghat drain had the maximum pollution load. These drains contained both storm water as well as effluents from septic tanks besides cases of open defecation. Dissolved oxygen was lowest for Krishnaghat drain and highest for Kurji drain water samples. BOD, total suspended matter, total suspended solids, ammonical nitrofen and total phosphates were highest for Krishnaghat drain. While chemical oxygen demand and chlorides were highest for Antaghat drain. The effect of discharge from various drains into the river Ganga at distances from 5 m onwards showed that the river tend to get more polluted because of the population of Patna (Table 33).

Table 32. Characteristics of major drains entering river Ganga at Patna (Rao et al, 1990).

Parameters	Drains				
	1	2	3	4	5
Temp.°C	16-33	20-32	21-33	20-30	20-30
pH value	6.9-7.8	6.6-8.0	6.8-7.6	6.0-7.4	6.4-7.3
DO mg/L	0.0-7.4	0.0-4.3	0.0-3.8	0.0-3.6	0.0-3.2
BOD_5 mg/L	6-32	6-65	18-30	45-140	35-170
COD mg/L	24-83	18-160	36-80	115-380	105-360
TS mg/L	520-840	540-780	510-800	550-970	720-1120
TSS mg/L	80-110	60-120	40-120	70-180	180-640
Chlorides mg/L	30-75	22-32	32-41	38-46	34-45
Ammtoniacal Nitrogen as N,mg/L	1-28	11-28	20-28	14-28	22-14
Total Phosphate as PO_4 mg/L	0-18.2	1.8-4.5	1.5-7.2	0.6-4.0	2.2-4.8

1. Kurji Drain 2. Rajapur Drain 3. Mandiri Drain 4. Antughat Drain 5. Krishnnghat Drain

Observations (Table 33) indicate that the characteristics of the effluents from M/S Bata India Limited and McDowell & Co Ltd both at Mokamah, which were not treated before discharge into the river. All these values were exceedingly high for McDowell effluents.

Table 33 . Effluent characteristics of two major industries at Patna (Rao et al, 1990).

Parameters	Industries	
	Bata	McDowell
1. pH	7.6	5.5
2. Total solids mg/L	4800	25600
3. Total suspended solids mg/L	1300	11000
4. Total volatile solids mg/L	1020	16400
5. BOD_5 at 20°C mg/L	440	18400
6. COD mg/L	1040	23600
7. Chlorides mg/L	2400	–
8. Hexavalent Chromium mg/L	ND	–

Ganga river water quality at Mokamah upstream and downstream (Table 34) show that five day BOD levels upto 100 mg/l were observed downstream 15 m. At this sampling site all values of physico-chemical parameters were highest except dissolved oxygen which was lowest. As the distance of sampling was increased upto 2 km the pollutant concentrations decreased,may be due to dilution factor.

Table 34. Ganga river water quality at Mokamah (Rao et al, 1990)

	Parameters						
Sampling points	pH	DO mg/L	BOD_5 mg/L	COD mg/L	TS mg/L	TSS mg/L	Cond msm^{-1}
Upstream, 200m	7.3	7.2	1	12	260	35	32.6
Downstream, 15m	6.3	1.4	100	240	980	90	53.3
Downstream, 100m	7.6	4.3	30	80	350	40	46.7
Downstream, 300m	7.6	7.1	1	12	265	32	33.9
Downstrcam, 500m	7.0	7.2	0.9	12	245	35	33.2
Downstream, 2 km	7.5	7.1	0.8	16	245	35	31.7

River Yamuna, the main source of water supply for Dehli is heavily polluted by sewage sludge and industrial wastes at several points along its 20 km stretch between Wazirabad and Okhla intakes. Out of the total quantity of water supplied in Delhi

20 per cent covers consumptive use, the remaining 80 per cent flows back into the river. There are about 17 open drains that discharge the wastes in Yamuna. The quantity of pollutants are dangerously high in seven drains namely - Najafgarh, Barapurah, Tughlakabad, trans-yamuna MCD, Sen Nursing home, Maharani Bagh, and Kalkaji. Out of them three viz., Najafgarh, Rajghat drain, and Barapurah drain greately contaminates the river near Delhi Gate I and Delhi Gate II. The Najafgarh drain contains high concentration of DDT and chlorohydrate. On the basis of the arbitary calculations the pollution contribution of the surveyed industries in the Najafgarh drain basin amounted to over 15,000 m^3/day of wastes effluents containing 6000 kg/day of fixed dissolved solids, 3000 kg/day heavy metals, 300 kg/day ink and dyes, and 200 kg/day detergents. It also received large volumes of extremely hot water from the power generator plants situated on the bank of this river. The industries that cause pollution are printing, electroplating, soap manufacture, food products, rubber, plastic, chemical, petrolium complex, fertilizer factories, synthetic material plants and drugs etc. Pollution of the Yamuna between Delhi and Agra (Table 35) has become a serious health hazard. Frequent power breakdown in Delhi bring the seven sewage treatment plants to a stand-still, resulting in dense sewage going into the river in place of treated one. The statement made by Prof. Arceivala that, 'Agra may be drinking Delhi's sewage' is wen documented by an analysis of the river along its 195 monitoring stations downstream to Agra, which revealed increasing daterioration in its quality (Chauhan, 1977). The Najafgarh drain came to be known as "sorrow of Delhi" after the famous Jaundice episode of Delhi in 1955, is still a "sorrow of Yamuna", as the river Yamuna continues to receive 3,18,180 m^3/day of sewage and industrial wastes from the drain (Chakrabarty and Gupta, 1980) At Agra Yamuna also receives the heavy load of city sewage and industrial effluents. Further in Mathura, on the bank of Yamuna an oil refinery has been set up. It has further deteriorated the quality of water of Yamuna.

Table 35. Biochemical characteristics of river Yamuna at different places in India.

Particulars	Delhi	Okhla	Mathura	Agra
BOD 5 days	2	5	9	12
Chlorine mg/l	18	46	111	140
Sulphates mg/l	29	44	110	112
Nitrates mg/l	0.8	1.3	2.1	3.8
Coliforms mpn/100 ml	150	24000	84000	240000
Entsrococci mpn/100 ml	21	211500	46000	150000

Direction of the water flow

Regùlar analysis of Yamuna water at two extremes of Delhi–Wazirabad (incoming) and Okhla (outgoing) showed deterioration in water quality with respect to chloride, ammonium nitrate permanganate value, coliform and faecal streptococci (Figure 15 and 16). The CPCB had monitored the Yamuna river water at 16 points from its source at Hatimkund to its confluence with the Ganga at Allahabad, had found the water quality worst at Delhi. It is not suitable even for bathing at ghats between Wazirabad and Okhla. Thus, water downstream of Wazirabad is quite polluted. This was due to addition of more than 225 million gallons per day of sullage besides industrial effluents of which only 49 mgd got full treatment and 52 mgd were partially treated.

In Madhya Pradesh important river (Table 36) suffer from severe industrial pollution. Das (1988) revealed that the highest quantity of effluents were discharged in Chambal and the lowest in Kshipra. The maximum river length affected was 96 km of river Khan, while the minimum was 6 km for Narmada and Betwa. The towns and villages affected by water pollution was largest (156) around Sone and smallest (4) around Betwa river.

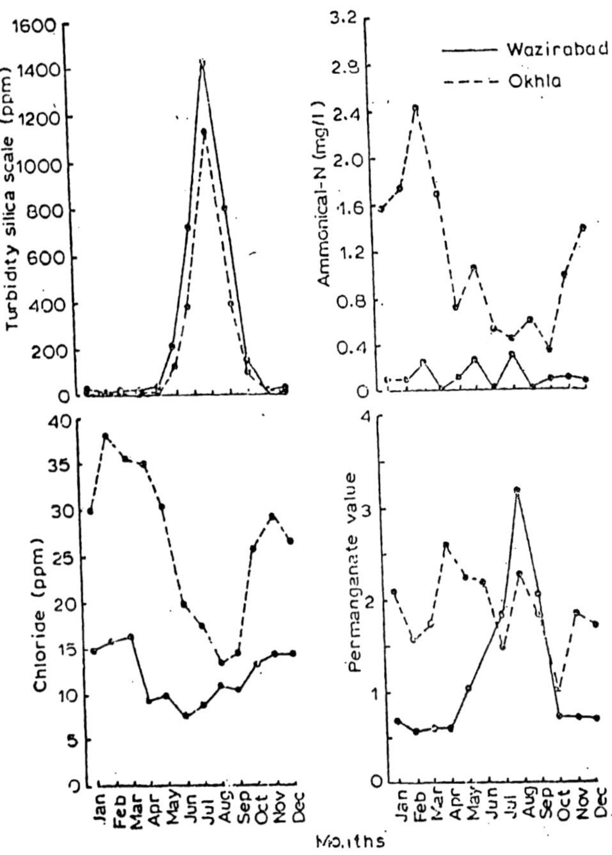

Figure 15. Seasonal variation in quality or Yamuna water due to discharge of effluent at Wazirabad and Okhla (Mean of 1970-71)

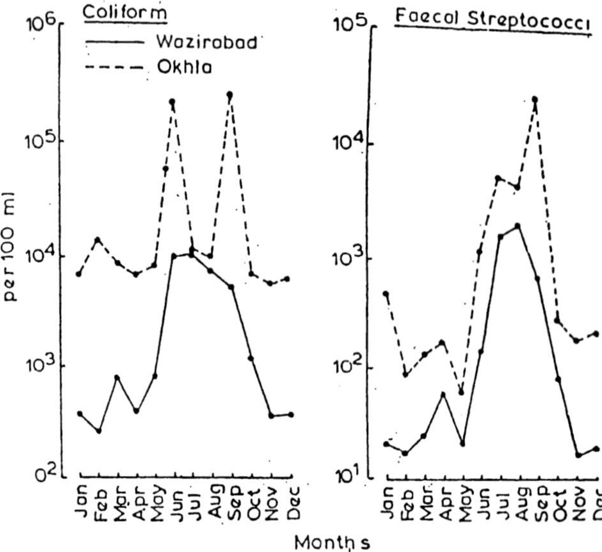

Figure 16 . Seasonal variation in bacteriological quality of Yamuna water due to discharge of effluents at Wazirabad and Okhla (1970-71 mean).

Table 36 . River pollution in the state of Madhya Pradesh (Das, 1988).

River	Industry	Effluent Quantity (M.G.D.)	River length affected (km.)	Towns and villages affected
1	2	3	4	5
Chambal	Rayon Mills, Nagda ; Caustic Soda Mills, Nagda	20	64	66
Kshipra	Textile Mills, Ujjain	1	8	10
Khan	Textile Mills, Indore	3	96	112
Narmada	Security Paper Mills, Hoshangabad	2	6	8

1	2	3	4	5
Betwa	Strawboard Mill, Vidisha	0.3	6	4
Tapti	Newsprint Paper Mills, Nepanagar	8	56	73
Sone	Orient Paper Mills, Amlai.	16	95	156
Indravati	Iron Ore Project Baila-dilla	4	42	40
Patra-Halali	Straw Products, Textile Mills and Distillery	2.5	24	20

Source: Water Pollution Control Board, Bhopal, Madhya Pradesh.

Khan river originates from Nimoli tank, about 10 km from Indore city and merges into Kshipra river near Ujjain. Pour major and several minor industries discharged their effluents directly into Khan river. Water samples collected from four sampling stations where major industries - Bhandari mill, Swadeshi mill, Kalyan mill and Malva mill release their effluents - showed high concentration of many industrial toxic pollutants (Table 37).

Contamination of Hardeo river occurs due to mergence of Korba thermal power station ashpond effluents (Table 38), The key parameters of the ash pond effluent were pH (8.20-8.35), total suspended solids (790-1080), total dissolved solids (620 mg/l), calcium (24.0-27.1 mg/l) and magnesium (9.6-9.7 mg/l). These parameters in Hardeo river waters exhibited abrupt rise in their values after mergence of the effluents into the river. At the confluence point of the effluents and river, the pH of the river water showed a significant increase from 7.40 to 8.31. The pH than decreased gradually at subsequent sampling sites. The distance between successive sampling sites of the river was about 1 km. The total suspended solids value at the confluence point increased to 1110 mg/1 which was more than 3 fold of the value in the river water before the mergence of the effluent stream. Similar increase in the value of other parameters such as total dissolved solids, Calcium and Magnesium was also noticed. Toxic metals such as Mercury, Lead, Arsenic, Manganese, Nickel, Cadmium and Zinc which were at undetectable levels in the river water in the upstream direction had appeared in it as a result of

Table 37. Physico-chemical features of industrially polluted water of Khan river (Shrivastava et al, 1987).

S. N.	Sampling stations	Temp. 0°C	pH	Suspended solids (ppm)	Dissolved solids (ppm)	D.O. (ppm)	COD. (ppm)	Average concentration (ppm)					
								Zn^{++}	Mn^{++}	Ni^{++}	Cr^{++}	PO_4^{--}	SO_4^{--}
1.	Bhandari Mill area	29	8.0	227.5	2001.75	6.51	136.4	2.8	0.4	1.4	2.1	1.764	95
2.	Swedeshi Mill area	30	8.4	268.40	1687.00	7.78	930.4	2.9	0.8	2.9	2.2	3.866	17
3.	Kalyan Mill area	33	7.7	471.00	3841.00	3.86	480.0	4.1	1.5	2.3	6.5	3.147	27
4.	Malwa Mill area	32	7.2	370.00	1078.00	3.65	428.3	4.5	0.5	0.3	0.9	4.764	108

Table 38. Physico-chemical characteristics of Hardeo river water due to discharge of Korba thermal power station ashpond effluents (Pervez and Pandey, 1994).

Parameters	Ash pond effluent stream Sample number				Unpolluted river water (upstream) Sample number				Polluted river water (down stream) Sample number			
	I	II	III	IV	I	II	III	IV	I	II	III	IV
pH	8.30	8.35	8.20	8.25	7.27	7.10	7.30	7.40	8.31	8.00	7.90	7.85
TS	1410.0	1610.0	1690.0	1700.0	510.0	610.0	390.0	410.0	1260.0	1002.0	990.0	980.0
TSS	790.0	990.0	1070.0	1080.0	435.4	535.5	309.7	330.6	1110.0	841.8	819.7	789.8
TDS	620.0	620.0	620.0	620.0	74.6	74.5	80.3	79.4	150.0	160.2	170.3	190.2
Ca	24.0	26.0	27.1	27.0	16.0	15.1	12.7	14.2	26.7	24.2	24.28	22.0
Mg	9.7	9.6	9.7	9.7	7.2	7.1	6.9	7.2	8.6	8.0	8.9	8.9
SO^-_4	0.3	0.3	0.3	0.3	0.1	0.1	0.1	0.1	0.4	0.5	0.5	0.5
Hg	0.4	0.4	0.4	0.4	NIL	NIL	NIL	NIL	0.4	0.3	0.4	0.4
Pb	6.0	6.0	6.0	6.0	NIL	NIL	NIL	NIL	6.0	5.0	5.0	6.0
As	0.5	0.5	0.5	0.5	NIL	NIL	NIL	NIL	0.5	0.5	0.5	0.6
Mn	20.0	21.0	20.0	21.0	NIL	1.0	1.0	NIL	18.0	17.1	18.5	19.0
Ni	2.0	1.0	1.0	1.0	NIL	NIL	NIL	NIL	1.0	NIL	NIL	NIL
Cd	4.0	4.0	4.0	4.0	NIL	NIL	NIL	NIL	1.0	2.0	1.0	1.0
Zn	9.0	9.0	9.0	9.0	NIL	NIL	NIL	NIL	4.0	5.0	7.0	8.0

All values except pH are in mg/L,**Nil denotes undetectable level.

the mergence of the effluent stream. The effect of the effluent was noticeable in the 3 km stretch of the river after mergence of the effluents.

Ratlam has a cluster of industries comprising Jayant Vitamins Ltd, Ratlam Alcohol and Carbon dioxide unit, Sajjan Impex and other small scale industries. The effluents from Alcohol unit and TLV effluents have moved through underground seepage to Kudel river about 8 km away and the coloured water is no longer potable. The water pollution is further increased with the effluents of Straw Board Mills, Sajjan mills and many small chemical industries.

Dewas has a Bank Note Press, Tata Exports, S.Kumars, Dewas synthetics etc. The effluents of these industries joins the Nagdhaman river, which ultimately joins Kshipra river.

Indore has seven textile mills, which discharge their effluents in Kshipra river.

Ujjain city itself in located on the bank of river Kshipra and has Shree Synthetics and ISCO pipe factory besides small scale industries like Hira mills, Binod mills, Rajdhani distillary, government milk dairy etc. The holy Kshipara receives the effluents of all these industries at Havankhedi.

The average of BOD and COD load from different industries (Table 39) shows that BOD was maximum (25000 mg/l) from Carbon dioxide unit at Ratlam and minimum (270 mg/l) from Shree synthetics at Ujjain. The COD was maximum (96000 mg/l) from Rajdhani distillary, Ujjain and minimum (831 mg/l) from Tata Exports Ltd, Dewas.

Table 39. The average load of BOD and COD from different industries (Shrotriya and Dubey, 1989).

Name the Industries		Parameters	
		BOD mg/l	COD mg/1
1		2	3
Shree Synthetics Ltd	Ujjain	270	1,496
Rajdhani Distillary	Ujjain	3.800	96,000
Sajjan Impex	Ratlam	450	6,700

1		2	3
Ratlam Alcohol and-			
Carbondioxide Plant	Ratlam	25,000	60.000
Jayant Vitamins Ltd.	Rallam	1,050	7,000
Tata Exports Ltd.	Dewas	386	831

The average pollutant load in Kshipra river through Nagdhaman (Dewas), Khan (Ujjain) and Kshipra (Ujjain) have been shown in Table 40, that in Chambal system at Nagda in Table 41, and that of Maleni system in Table 42.

Table 40. Average pollutant load (mg/1) in Kshipra system (Shrotriya and Dubey, 1999).

Parameters	Subsystems of Kshipra River		
	Nagdhaman Dewas	Khan Ujjain	Kshipra Ujjain
T.D.S. mg/l	411	389	326
T.S.S. mg/l	174	99	92
BOD mg/l	104	5	4
COD mg/l	176	30	30
CL^- mg/l	58	87	44
$S04^-$ mg/l	351	367	202
Total Alk. mg/l	-	74	59
Sodium mg/l	52	46	28

T.D.S. - Total dissolved solids, T.S.S. - Total suspended solids BOD-Biochemical oxygen demand, COD-Chemical oxygen demand, CL-Chlorides. SO4-Sulphates, Total Alk,-Total Alkalinity.

Table 41. Average pollutant load (mg/1) in Chambal system at Nagda (Shrotriya and Dubey, 1989).

Parameters	Without Treatment 1976-77	Partial Treatment 1981-82	Full Treatment 1985-86
1	2	3	4
T.D.S.	1,133	750	500
T.S.S.	184	72	50

1	2	3	4
BOD	80	65	45
COD	193	160	95
CL^-	138	81	60
SO_4^-	318	236	600
Total Alk.	20	13	40
Sodium	214	57	70

Table 42. Average pollutant load (mg/1) in Maleni system (Shrotriya and Dubey, 1989).

Parameters	Jadvasa Nallah	Kurel	Maleni
T.D.S.	49	29	418
T.S.S.	43	33	92
BOD	1,215	135	05
COD	1.959	216	45
CL^-	4,453	315	129
SO_4^-	10,269	2,866	1,517
Sodium	124	71	80

The Shivshakti paper mill is situated near the bank of river Betwa in Vidisha (MP). The mill has no proper effluent treatment unit and its effluent was drained into the Betwa river after partial settling only. The mill used wheat straw as raw material and manufactures brown paper. The physico-chemical characteristics of diluent water, effluent and dilutions (Table 43) indicates that even 20 percent diluted effluent, and 20 percent diluted effluent after aeration were highly polluted. The discharge of such polluted effluents were deteriorating the water quality of river Betwa.

Table 43. Physico-chemical characteristics of diluent water, effluent and effluent dilutions Shivshakti paper mill in Vidisha (Chouhan et al, 1990).

Physico-chemical charecteristics	Diluent water	Undiluted effluent	20% Diluted effluent	20% Diluted effluent after aeration
Water				
Terrperature (°C)	30	23	26	26
Conductivity (μmho)	0.24	191	1.88	1.88
Turbidity (ppm)	10	70	38	38
Dissolved				
oxygen (mg/l)	12.0	2.0	3.2	12.8
Free CO_2 (mg/l)	1.1	145.2	118.8	57.2
Alkalinity (mg/l)	175	30	520	520
Hardness (mg/l)	167	280	240	240
Chlorides (mg/l)	0.71	132.06	119.28	119.28
pH	8	5.5	6.5	6.5

The mean values of physico-chemical characteristics of the industrial effluents of Birla Nagar, Gwalior (Table 44) show that they were dark green to black in colour. The BOD and COD values were quite high. On comparison with those of the standard set by ISI, it was found that the waste water from industrial complex at Birla Nagar was not fit for its disposal on land in the present form as the waste water is invariably coloured and possess high BOD, COD and total solids. Though, this wastewater is currently being disposed on the land and used for irrigation without treatment, it is suggested that it should be disposed or used for irrigation after proper treatment.

There is no adequate sewerage system in the cities of Madhya Pradesh. Indore, the largest city of the state is served by limited sewer lines since 1928. The rest is served by the old methods of manual collection of night soil and its disposal in trenches. The worst example in that of Jabalpur city where sewage out of censervancy latrines is allowed to flow through open surface drains and is ultimately discharged into Omtinala which flows through the densely populated parts of the city (Shrivastava, 1987). River pollution caused due to municipal waste waters in six major cities of Madhya Pradesh (Table 45) has also been

Table 44. Physico-chemical characteristics of textile industry effluents at Birla Nagar, Gwalior (Mishra and Saksena, 1990).

S. No.	Characteristics	Effluent at Birla Nagar	Tolerance limit for industrial effluents		
			Into inland surface waters	Into public sewers	On land for irrigation
1.	Colour	Dark variable	Colourless	Colourless	Colourless
2.	Suspended solids, mg/l^{-1}	476.80	100	600	200
3.	Dissolved solids, $mg.l^{-1}$	933.86	2100	2100	2100
4.	pH value	8.04	5.5 to 9.0	5.5 to 9.0	5.5 to 9.0
5.	Temperature, °C	26.91	shall not exceed 40 in any section of the stream within 15 meters down strem from the effluent outlet	46 at the point of discharge	–
6.	BOD, $mg.l\text{-}^{1}$	298.95	30	350	100
7.	COD, $mg.l\text{-}^{1}$	542.48	250	–	–
8.	Chlorides, $mg.l^{-1}$	74.34	1000	1000	600
9.	Dissolved phosphates $mg.l^{-1}$ (as P)	0.015	5	–	–

estimated by Das (1988). It is heartening to note that the holy Kshipra river receives sewage of UJjain and Indore cities.

Table 45. River pollution by municipal waste water of major cities in the state of Madhya Pradesh (Das, 1988).

City	Quantity (MGD)	Treatment	River drain being Polluted	No. of villages affected
Indore	7	Partial+Sewerage farming	Khan	112
Ujjain	4	Partial	Kshipra	10
Bhopal	14	8 MGD Treated + farming	Patra and Halali	20
Jabalpur	12	No treatment	Local drains	–
Raipur	6	No treatment	Local drains	–
Gwalior	10	Land irrigation	Local drains	–

Source : Water Pollution Control Board, Bhopal, Madhya Pradesh.

Table 46. River pollution in Madhya Pradesh (Das, 1988)

River	Industry	Effluent quantity (MGD)	River length affected (Km)	Towns and villages affected
1	2	3	4	5
Chambal	Rayon mills, Nagda Caustic soda mills, Nagda.	20	64	66
Kshipra	Textile mills,Ujjain	1	8	10
Khan	Textile mills, Indore	3	96	112
Narmada	Security paper mill, Hoshangabad	2	6	8
Betwa	Strawboard mill, Vidisha	0.3	6	4
Tapti	Newsprint paper mill Nepanagar	8	56	73

1	2	3	4	5
Sone	Orient paper mill, Amlai	16	95	156
Indravati	Iron ore project, Bailadilla	4	42	40
Patra-Halali	Straw-products Textile mills, and Distillary	2.5	24	20

In Madhya Pradesh river Sone receives 14 million gallons/day of wastes from Orient Paper mills, limited at Amlai which has pollutted 60 km stretch of the river from the point of waste discharge. The Shahdol town, which is about 40 km downstream from the factory discharge, is facing the problem of finding alternate source of water supply. In the second zone i.e. Bihar this river receives 40.85 x 10^4 gallons/day on industrial wastes from Pulp and Paper, Chemicals, Vanaspati and Cement industries at Dalmianagar. The blakish brown highly organic and alkaline wastes contain a number of chemical substances like merceptans, sulphides, resin, soap and free chlorine which deteriorate the physico-chemical quality of the river water by a number of ways.

The high organic content of the waste stress heavily on the oxygen budget of the stream and the water shows anaerobic condition in more than 24 km stretch of the river.

River Khan receives about 6.0 million gallons/day of sewage and 3.0 gallons of textile waste giving a bad small up to 24 km from Indore. With this mixture, the river joins Kshipra river near Ujjain, and the place of confluence is known as Triveni Sangam. Due to high amounts of phosphate at Triveni Sangam, frequent formation of algal blooms in Kshipra is reported, causing innumerable problems to Ujjain Water Works.

River Tapti receives about 6.0 to 8.0 million gallons/day of untreated waste from National News Print and Paper Mills Limited Napanagar. The water quality of Tapti river was found to be deadly affected in summer months even upto a distance of 32 km from the point of waste discharge.

River Chambal receives 20 million gallons/day untreated waste from Gwalior Rayons and Silk Industry at Nagda in Madhya Pradesh. A number of surveys have indicated that pollution produces striking changes in the biotic comiminity. Some species may be unable to survive and others may persist in reduced coactions, certain species may be able to attain greater abundance. These structural changes are very useful in assessment of water quality. Trivedi (1980, 1981) made an extensive study on the Chambal river polluted by Rayon and Caustic soda industrial complex at Nagda on Benthos (Table 47) and showed that the diversity index(d) can be correlated with various degrees of pollution (Table 48).

Table 47. Correlation between Shannon and Weaver's diversity index (a) and important pollution parameters in the Chambal river at Nagda (Trivedi 1981).

Period	Stations	pH	DO mg/1	BOD mp/1	TDS mg/1	Zn mg/1	$\bar{d}$
1	2	3	4	5	6	7	8
Nov.	I	8.20	7.61	5.90	258	Trace	4.63
Dec 1976	II	7.21	2.90	36.8	632	1.53	2.77
	III	7.52	3.88	27.8	532	0.04	3.31
	IV	7.75	4.81	24.4	466	3.84	5.60
	V	7.99	5.58	21.4	419	Trace	4.17
	VI	8.11	5.81	19.2	381	Trace	4.52
Jan	I	8.00	8.20	7.2	272	Trace	4.57
Feb 1977	II	4.26	0.50	59.5	906	2.55	2.33
	III	5.00	2.20	42.9	702	1.81	2.65
	IV	5.60	2.91	34.1	597	1.34	3.33
	V	6.32	3.62	28.8	526	0.46	3.40
	VI	7.29	4.81	24.7	462	Trace	4.17
Mar	I	8.40	6.5	11.3	310	Trace	4.82
April 1977	II	3.92	NIL	154.5	2039	7.23	1.44
	III	4.86	NIL	105.7	1540	6.38	2.11
	IV	5.72	0.5	82.2	1229	5.48	2.72
	V	6.50	1.8	68.9	1040	4.25	2.82

1	2	3	4	5	6	7	8
	VI	7.21	2.5	61.2	921	2.70	3.39
May	I	–	–	–	–	–	–
Jun 1977	II	2.21	NIL	260	3251	13.50	0.0
	III	3.25	NIL	163	2440	11.55	0.57
	IV	3.98	NIL	120	1.981	8.48	0.75
	V	4.76	NIL	97	1663	5.45	0.89
	VI	5.18	0.5	82	1487	3.87	1.74

Table 48. Species diversity of Banthos invertebrates (d) in Chambal river at Nagda receiving pollution from Rayon and Caustic effluents (Trivedi, 1980).

Station	Above outfall	Below outfall	Down stream	
I	4.63	2.77	3.83	4.2
II	4.56	2.33	3.33	4.18
III	4.82	1.44	2.72	3.39
IV	-	0.00	0.75	1.74

In Uttar Pradesh, Western Kali, Eastern Kali, Hindon and Krishna rivers have been studied. The western Kali receives about 4.2 x 10^3 m^3 /day of wastes from Sugar factory and Distillary at Mensurpur. Eastern Kali receives about 145 x 10^3 m^3/day from Sugar factory at Khatauli, Daurala Sugar distillary and Chemicals, Daurala, Modi Rubber, Modipuram, and Panniji Sugar factory, Bulandshahr along with the city sewage of Meerut and Bulandshahr. Hindon receives about 14 x 10^{-3} m^3 /day wastes from Pulp and Paper factory Shahranpur, and Krishna river receives about 7.2 x 10^3 m^3/day wastes from Sugar factory and distillary at Shamli. Due to discharge of large volumes of organic wastes, long stretch of these rivers has developed an anaerobic condition, making the water unfit for use. Eastern Kali river in Meerut is considerably sporadically polluted from waste industrial effluents, particularly with respect to Vanadium and Chromium (Kudesia and Sharma, 1981).

River Gomti along its entire stretch receives the maximum pollution load at Lucknow. 33 waste outfalls, bringing about 21.26

million gallons of wastes per day from Pulp and Paper factory, distillary and sewage, badly deteriorate the water quality of this river. Calculations for self-purification capacity indicates that even after travelling a distance of 65 km downstream of Lucknow, the recovery of water is not complete.

The river Terhi in Gonda district of Uttar Pradesh, is connected with a sugar and distillary industry's effluent drain at Nawabganj industrial area. Observations (Table 49) shows the physico-chemical conditions of the river water. The apstream had desirable parameters. However, downstream where industrial effluents mixed with river water, the values of all physico-chemical parameters were high enough and indicate a fairly good amount of organic pollution. The high ammonia content might be due to entry of washings from the drainage of the town and adjacent area.

Table 49. Physico-chemical conditions mg/l of river Terhi at Nawabganj industrial area in Gonda district of UP (Pande and Patel, 1997).

Parameters	Upstream		Downstream	
	Range	Mean	Range	Mean
Temperature (°C)	16.0-32.0	24.83	16.0-35.0	25.75
pH	7.0-7.2	7.05	6.5- 7.5	6.96
Dissolved Oxygen	10.0-17.0	12.91	1.5-6.0	3.67
Free CO_2	5.5-3.0	4.08	2.0-4.8	3.21
BOD (20°C, 5 days)	4.0-5.5	4.59	1680-18635	7236.22
COD	10.2-22.5	14.63	935.0-2235.0	1387.62
Phosphate	0.52-2.5	1.61	2.4- 4.6	3.22
Total Nitrogen	2.0-6.5	4.70	17.3-52.0	31.61
Ammonia	Nil-0.56	0.096	0.60-13.5	6.01
Calcium	18.0-34.0	24.5	56.0-360.0	158.08

Mathura has about 300 Saree printing cottage industries, which are contributing significantly to water pollution, by releasing enormous quantity of coloured effluent. Observations (Table 50) reveals that these effluents have higher values nearly for all the parameters except dissolved oxygen. The effluents were loaded with low dssolved oxygen (5.4 mg/l) but very high Chemical Oxygen Demand (688 mg/l). Higher COD in effluent indicates that it needs more oxygen to stabilize organic substances like dyes and starch for rapid oxidative degradation, for which it will consume dissolved oxygen.

Table 50. Physico-chemical characteristics of Saree printing industry effluent at Mathura (Jain and Agarwal, 1989; Jain and Kumari, 1990).

Parameters	Unit	Characteristics
Colour	Visual	Purple red
PH		8.2
Specific conductance	µmhos/gm.	1960
Calcium as Ca	mg/Litre	196
Magnesium as Mg	mg/Litre	84
Bicarbonate as $CaCO_3$	mg/Litre	340
Chloride as Cl	mg/Litre	500
Total alkalinity	mg/Litre	350
Total hardness	mg/Litre	850
Dissolved oxygen	mg/Litre	5.5
B.O.D.	mg/Litre	97.0
C.O.D.	mg/Litre	690

The effluents of Mathura Petroleum refinery were analysed for their physico-chemical characteristics. Samples were taken from four different sites: First, just at the outlay of the refinery; second, about 200 meters downstream; third, where the stream joined the Yamuna canal; and fourth, about 1 km down the canal. The physical characteristics (Table 51) reveal that the effluents were alkaline in nature. The high pH may be due to the presence of malts in high concentration. These results are supported by high conductivity of the effluents. The phenolic content was also fairly high at site 1 and 2. It was however, lower at site 3 and 4. The fall in the concentration of the phenolic content was due to the dilution of the effluents by the canal water which was used for the irrigation.

Perusal of data (Table 52) indicate that the most dominating phenol in the effluents was 0-Chlorophenol which was about 17.03 mg/l (26%) of the total phenolic content. Other dominating phenolic compound were p-cresol 13.84 mg/l (21.3%), and 2,6-dichlorophenol 13.33 mg/l (20.51%). The overall assessment indicate that the effluents carry considerable amount of phenol and make the receiving water unfit for drinking. The ground seepage of such effluents may deteriorate the quality of ground water.

Table. 51. Major Characteristics of the. refinery Effluents. (Gupta, 1989).

Parameters	Sampling Site			
	1st	2nd	3rd	4th
Temperature in °C	22.5	22.5	22.1	22.0
pH	8.2	8.2	7.9	7.9
Conductivity in m.mhos	1.68	1.65	1.65	1.40
Total phenols as C_6H_5OH in mg/litre	65	62	32	30

Table 52. Qualitative analysis of the four effluents of refinery by GLC. (Gupta, 1989).

S. No.	Name of the compound	Boiling point °C	Weight of compound in mg/litre			
			1st	2nd	3rd	4th
1.	O-chlorophenol	176	17.03	16.31	08.42	07.89
2.	Phenol	182	05.26	04.71	02.39	02.25
3.	O-cresol	192	08.07	07.75	04.01	03.76
4.	m-cresol	201.8	01.76	01.74	00.95	00.88
5.	p-cresol	202.4	13.84	13.35	06.92	06.49
6.	2,6-dichlorophenol	206	13.33	12.82	06.56	06.14
7.	p-chlorophenol	217	05.73	05.28	02.72	02.55

River Suvaon, a tributary of Rapti, receives 76 x 10^4 gallons per day of wastes from Balrampur Sugar factory, which creates anaerobic conditions. The anaerobic conditions with other factors like redox potential and constant supply of sulphate with the waste, stimulate activation of sulphur reducing bacteria producing hydrogen sulphide, a toxic gas and bad smalling gas in the river.

River Subarnarekha, a main source of water supply for Ranchi and Jamshedpur receives the wastes from Ranchi distillary and the Sewage from Ranchi and Jamshedpur township. The BOD downstream of the distillary waste goes upto 104 mg/litre during the month of May it is about 0.3 mg/litre.

In Rajasthan at Kota, the Chambal river receives city sewage without any treatment, which pollute the water and make it unfit for plankton and fish life, which is a source of food for the people of this region. It is also an eye sore for the Railways whose intake is just near the Railway Bridge, to provide drinking water to Railway Colony and the Railway station. This river also receives the wastes from industries like Shriram Chemicals, Shriram Rayons, J.K.Synthetics etc., at Manas village. Below the waste discharge the river water detriorates so much so that it is not suitable for domestic irrigation and fishing in the next 32 km stretch of the river. Besides city sewage and industrial effluents, insecticides and fertilizers sprayed in the fields, and heavy amounts of detergents from domestic use also reach the river by surface run-off.

Thus Chambal river waters are in a state of flux and the addition of untreated effluents, sewage, various kinds of organic and inorganic compounds contributes a major degree of pollution of this river.

In Rajasthan Agarwal (1983) studied the water quality of three main drains entering river Chambal at Kota (Table 53). It was observed that chlorine concentration in all the three drains was high, and they belonged to strong category of Klein (1957) and suggested their pollution by organic matter (Adoni, 1975). The high ammonical nitrogen in comparison to nitrate nitrogen was due to discharge of untreated sewage, with incomplete oxidation. The presence of calcium and magnesium results in essentially calcarious hard waters.

Table 53. Chemical characteristics of sewage drains entering river Chambal at Kota (Agarwal, 1983).

Characteristics	Chandraghata	Kunadi	Rampura
Chlorine (ppm)	80-600	78-400	62-580
Ammonical nitrogen (ppm)	7.5-13.3	5.3-12.6	3.3-10.2
Nitrate nitrogen(ppm)	0.031-0.163	0.015-0.182	0.028-0.280
Calcium (ppm)	6.8-15.2	2.3-9.0	3.4-11.6
Magnesium (ppm)	9.2-16.3	4.6-15.5	7.2-13.0

Observations on physico-chemical characteristics of Chambal river water (Table 54) indicate that:

1. The values of transparency, pH, dissolved oxygen, and dissolved organic matter were low in summer and high in winter.
2. The concentration of bicarbonates, phosphate, silicate, total nitrogen, chlorine and total dissolved solids were higher in summers and lower in winters.
3. High values of chlorine, nitrogen, phosphate and organic matter in the water of this river suggested its pollution by organic matter chiefly of animal origin.
4. The data on dissolved oxygen and dissolved organic matter indicate that oxidation was incomplete during the winter season.

Table 54. Physico-chemical characteristics of Chambal river water at Kota (Agarwal, 1986)

Parameter	Trans-perency (cm)	DO_2 (ppm)	Phos-phate (ppm)	Total Nitro-gen (ppm)	Chlo-rine (ppm)	Dissolved organic mattes (ppm)
1	2	3	4	5	6	7
January 83	30.6	16.5	0.28	5.7	13.4	11.6
February	38.5	13.2	0.23	6.8	14.6	13.4

1	2	3	4	5	6	7
March	25.0	11.2	0.20	7.2	24.5	10.8
April	15.8	10.0	0.30	8.6	22.8	10.4
May	10.5	8.4	0.37	11.3	21.3	8.5
June	9.8	7.8	0.42	14.7	28.5	6.8
July	8.6	8.1	0.56	14.2	41.8	4.5
August	8.2	9.8	0.58	12.0	38.1	5.4
September	8.0	12.4	0.46	9.3	28.2	8. 6
October	8.3	12.2	0.47	8.0	17.1	9.2
November	22.2	14.7	0.41	7.6	16.4	12.6
December	25.5	16.0	0.35	6.0	14.8	9.2

Kota Saree printing industry effluents were slightly alkaline to highly alkaline (Table 55). Electrical conductivity was 25-31 μ moha/cm. Total suspended solids and total dissolved solids were higher than the maximum permissible limit for surface water. These higher values might be due to higher concentration of dyes in the effluents. Chlorine concentration was high. BOD was higher. COD was higher than BOD. From the physico-chemical analysis, it is evident that Kota Saree printing effluents were loaded with high pollution potential and as such need proper treatment before discharging them into streams or for land application.

Table 55. Physico-chemical characteristics of Kota Saree Printing effluents.

Parameters		Unit	Maximum Permissible limit	Saree Printing
1	2	3	4	5
1.	Colour	Visual	—	Dark/red
2.	Odour	Smell	—	Pungent
3.	Temperature	°C	40	28-35
4.	pH	—	5.5-9.0	8.7-11.2

1	2	3	4	5
5.	Total suspended solids	mg/l	100	185-284
6.	Total dissolved solids	mg/l	2100	8451-9868
7.	Electrical conductivity	μ mohs/cm	—	25-31
8.	Chloride	mg/l	1000	175-205
9.	Residual chlorine	mg/l	1.0	ND
10.	Nitrate nitrogen	mg/l	50	41-82
11.	Dissolved oxygen	mg/l	—	ND
12.	BOD (5 days at 20°C)	mg/l	30	140-285
13.	COD	mg/l	250	232-295

ND = Not detected.

Observations (Table 56) on physico-chemical characteristics of dying and textile effluents from river Bandi, district Pali in Rajasthan showed that the effluent was highly rich in chlorine, bicarbonates and total dissolved solids, The values of pH, BOD and COD were also not correlating with the tolerance limits.

Table 56. Physico-chemical characteristics of dyeing and textile effluent from river Bandi, district Pali (Dayama, 1987).

Parameters	Characteristics
Colour	Reddish
Odour	Pungent
Temperature °C	32.2
pH	10.2
Total solids	40,000 mg/Litre
Dissolved oxygen	Nil
Carbonate	10 mg./Litre
Bicarbonate	94 mg./Litre
Chloride	2436 ppm
Phosphate	7.2 ppm
COD	362 ppm
BOD	506 ppm

The textile industries in Pali number more than 500 and are largely confined to Pali proper. In the absence of proper effluent treatment from these industries, the environment of the city remains polluted. The discharges run through open channels in the city, and it is common sight to find stagnant pools of coloured water/foul water in the industrial pockets. Raw effluents finally enter into the nearby Bandi river and finally the river merge in Luni river. The effluents from these industries are not only biodegradable and are highly toxic. They contain Caustic soda/ soda ash, bleaching powder/hydrogen peroxide, chlorides, sulphates, acetic acid/Sodium acetate, Hydrosulphite of soda, Sodium nitrate, detergents, napthols, vat dyes, colours and pigments. Toxic metals like Copper, Zinc, Aluminium, Nickel, Chromium etc. from dyes and pigments used in the industry also form a part of the composition of the industrial wastes. The results of chemical analysis of raw effluents for individual processes as well as mixed samples of various operations used in a textile industry (Table 57) shows brief idea about the quality of these effluents.

Table 57. Observations of chemical analysis of grab samples of raw effluents from some textile industries of Pali (Mauria, 1986)

Process	Colour	pH	Suspended solids mg/l	Oil and grease mg/l	COD mg/l	BOD mg/l
Desizing	Creamyellow dirty turbid	12.80	1280.0	162.0	760.2	398.0
Bleaching	Cream yellow turbid	12.50	1185.0	142.0	818.7	547.0
Dyeing	Deep green turbid	9.85	395.0	36.0	833.9	657.8
Printing	Brick red turbid	8.40	235.0	28.0	672.7	474.0
Finishing	Milky white turbid	7.50	680.0	64.0	710.7	476.4
Composite effluent (1+3+4)	Dirty green turbid	12.50	525.0	81.0	935.6	571.8
Caustic treatment +2+4	Purple Fluorescent turbid	12.55	495.0	94.0	608.1	383.5

These effluents make the potable water polluted far beyond the lowest requirement for human consumability in Pali.

A single Super Phosphate fertilizer plant at Alwar is polluting 57 villages of Navkhada Bundh, Sikari Bundh and Ruparel river. The Soda ash plant of Modi Chemicals and Alkalies limited, will increase the existing pollution 100 times. Besides there are other 30 industries causing water pollution.

Water pollution due to effluents discharged from Copper Complex at Khetrinagar in the Jhunjhunu district of Rajasthan has been carried out by Joshi, Thukral and Chand (1982). Copper concentration in the tailing water from different spots increases in the water: discharge point< tailing pond< nullah. Well water has quite low concentration of copper (0.02 mg/l). Cobalt concentration is highest in tailing water from the discharge point (0.195 mg/l) and it was absent in well water. Manganese concentration was quite high in tailing water from discharge point (0.176 mg/l) and from nullah (0.151 mg/l). Even well water contained 0.125 mg/l of Manganese (Table 58). Zinc was absent in well waters. A gradual increase in the concentration of various metal in water occurs in the direction of its flow, probably due to high rate of evaporation. Storage of tailing pond water may lead to the contamination of water table in the vicinity. Tailing water is not suitable for irrigation because it will increase the heavy metal concentration in soil over the longer periods.

Table 58. Heavy metal (mg/l) in the effluents of copper complex, Khetrinagar, Rajasthan (Joshi, Thukral and Chand, 1982).

Metal	Discharge point	Tailing pond	Nullah	Wells
Copper	0.136	0.178	0.196	0.02
Cobalt	0.195	0.081	0.15	0.001
Manganese	0.176	0.05	0.151	0.125
Zinc	0.116	0.085	0.013	–

River Ib, receives a large volume of highly putriscible liquid wastes from the Orient Paper Mill, which creates a serious pollution problem at Brajrajnagar in Orissa, particularly during the summer, when the entire water from the river is drawn for use of the factory and no water is left for dilution purposes. This results in complete deoxygenation of the river water for atleast 3 km down stream. The low oxygen condition have been reported to affect the aquatic life for about 25 km down stream.

River Bhadra at Bhadravati (Mysore) receives 3818 gallons per day of untreated wastes from Mysore Pulp and Paper Mills and about 10,000 gallons/day from steel plant. As a result of which 7.55 to 19 ppm of chlorides and 0.82 to 12.2 ppm of sulphide have been detected even upto 18 km down stream from Bhadra confluence into Tunga river. These toxic substances cover a 13 to 14 km long river stretch below the outfall during the pre monsoon periods.

River Cauvery, the major source of drinking water in Tamil Nadu, caused endemic water borne diseases in many towns because of the addition of domestic and industrial wastes enroute (Pachpurkar, 1976; Paramasivan and Sreenivasan, 1981). The Seshasayee Pulp and Boards Limited, Pallipalayam which is situated on the bank of river Cauvery, manufactured all varieties of writing and printing paper, kraft etc., with a capacity of 65 tonnes per day. Bamboo was used for making paper. The wastes from the process plant were 6.5 - 70 Million gallons (30,300 to 32,000 m^3) per day. The composition of the sample at the outfall varied from time to time depending on the waste recovery and treatment. The effluents were invariably dark brown in colour. The chief characteristic of this effluent was its high BOD varying from 100 to 1800 mg/l. The pH was always alkaline except once. The high BOD had a little impact on the oxygen regime down stream for about 0.8 km, but there was no drastic reduction in dissolved oxygen. At certain points, a lot of fine pulp fibres settled down and sludge to a depth of about one meter was formed on the river bed. This was evident from the high suspended solids content of the effluents varying from 150 - 862 mg/l. The change in dissolved solids was reflected in the electrical conductivity of the water (Table 59).

Table 59. Hydrobiological conditions of river Cauvery (Sreenivasan and Sounderraj, 1967)

Parameters	Above the outfall	At the outfall	0.8 km below outfall
pH	7.5-8.7	5.9-9.6	7.8-9.1
Colour	Colourless	Brown	Colourless
Conductivity (μ mhos)	136-560	960-2100	200-400
Alkalinity (Methyl orange) (mg/l)	96-286	60-192	62-284
Total solids (mg/1)	54-72	294-1896	128-182
BOD (mg/1)	0.4-5.8	103-1800	10-250
Dissolved oxygen (mg/1)	4.2-9.2	0.0-3.0	2.3-8.4

River Kali around Dandeli, Karnataka, was contaminated with effluents from Perro-Alloy's Ltd, Plastic factory, Plywood factory and city waste near Dandeli New Bridge. physico-chemical parameters of river water (Table 60) show that the river water was alkaline with pH ranging from 7.6 - 7.8, Dissolved Oxygen was high upstream, but was low downstream near Dandellappa Swamy temple site, at this site dissolved organic matter, cations like Potassium, Magnesium, Calcium, Sodium and anions like carbonate, bicabonate, chlorides, Sulphates, phosphates, total hardness and silica were recorded in higher concentrations. Oxidised Nitrogen was high in site S_2, may be due to drainage from paddy fields. At S4 station there was dilution in all the parameters as compared to S 3 station.

Table 60. Physico-chemical parameters of Kali river water contaminated with Ferro Alloy's Ltd, Plastic factory. Plywood factory and city waste near Dandeli New Bridge (Bharati and Murthy, 1990).

Parameters	Upstream		Downstream	
	Station S_1 Near Moulangi	Station S_2 Near New Bridge	Station S_3 Near Dande-Happa Swamy temple	Station S_4 Near Kervada
Air Temperature °C	29.3	20.7	26.1	23.9
Water Temperature °C	26.4	25.8	25.5	25.4
pH	7.7	7.5	7.7	7.7
Dissolved Oxygen	7.8	7.8	4.1	7.0
Chemical Oxygen demand	13.8	17.4	113.8	37.2
Free-Carbon dioxide	3.2	2.6	6.1	2.1
Total Solid	75.0	79.5	369.5	07.0
Dissolved Organic matter	1.4	2.1	27.3	3.7
Carbonate	–	–	36.0	18.0
Bicarbonate	97.6	97.6	245.2	118.4
Chloride	18.1	20.4	83.4	36.0
Sulphate	3.2	3.0	96.6	16.8
Phosphate	0.015	0.030	0.070	0.052
Silica	4.0	6.0	12.0	8.0
Ammonia - Nitrogen, (NH_3)	0.373	0.222	0.37	0.31
Oxidized Nitrogen ($NO_3 + NO_2$)	0.560	2.184	0.896	0.604
Organic Nitrogen (O-N)	0.341	0.408	0.566	0.500
Total Hardness as $CaCO_3$	19.0	22.0	45.0	43.7
Calcium Hardness				
(i) As $CaCO_3$	14.5	16.5	126.9	30.8
(ii) As Ca	6.0	6.6	60.9	11.8
Magnesium	1.0	1.1	5.1	2.6
Sodium	4.6	5.2	192.0	96.0
Potassium	1.0	1.1	37.2	24.8
Iron	1.8	2.3	3.65	1.9
Manganese	3.0	4.0	4.5	4.2

All values except temperature and pH are in mg/L

Pykara river in the Western Ghats has Protein Products of India (PPI) industry on its banks, produce gelatin, ossein,and dicalcium phosphate by acid extraction of animal bones. The effluents produced are disposed into the Pykara river. The effluents physico-chemical parameters (Table 61) indicate the extent of pollution load exerted on the hydrobiology of Pykara river water. The effluent was turbid, brown coloured with obnoxious foul odour, pH varying from acidic to alkaline, BOD and COD values ranged from 140-190 and 1200-1840 mg/l respectively. The chloride, sulphate, ammonia, nitrate and nitrite were observed in the effluent in considerable quantities. Presence of phenolics,

Table 61 . Physico-chemical properties of effluents of protein industry, discharged in Pykara river (Rajnnan and Kandasamy, 1990).

Character	PPI effluent	PPI - Mixing point effluent	
		Sample I	Sample II
pH	6.5-7.8	7.8	5.5
Odour	Foul Odour	Foul Odour	Foul Odour
Colour	Brownish	Brownish	Brownish
Total suspended solids	160	140	190
Total dissolved solids	1840	1700	1200
Dissolved Oxygen	1.8	3.1	1.0
BOD	650	520	720
COD	1310	1150	1260
Chloride	500	320	800
Sulphate	340	210	400
Organic Carbon (%)	0.19	0.18	2.01
Total Phenols	2.8	2.1	3.2
NH_3-nitrogen	12.5	6.2	20.5
NO_2-nitrogen	15.2	12.6	17.2
NO_3-nitrogen	16.8	14.2	17.4
Phosphate	11.4	9.2	14.8

Values are expressed as mg/L, except for pH and organic carbon.
Station I=point of discharge; Stationll=downstream 100 m.

ammonia and other organic compounds, which came out as waste from bones after extraction of proteins were responsible for the foul smell. The suspended solids contributed to turbidity in the receiving water. The dissolved solids including chloride and sulphate were very high. Presence of low dissolved oxygen facilitate conversion of sulphate to sulphide which was toxic to fish. Ammonia in the effluent was also lethal to fish population. Nitrates, nitrites and phosphate quantity swell on accumulation over time. The deleterious effects of PPI effluents on fish *Tilalia mossambica* could be attributed to the low levels of dissolved oxygen, high biological oxygen demand and Chemical oxygen demand and other intricate compounds in the effluents.

River Bhavarri in the vicinity of South India Viscose limited at Sirumugai was polluted with acidic effluents. Owing to unsteady lime treatment of the effluent, some times acidity persisted and the affluents has free mineral acid as high as 280 mg/l. The fall in dissolved oxygen content was greater because of the presence of free mineral acid or calcium hydroxide or sodium sulphate (Table 62).

Table 62. Hyrdrobiological conditions of river Bhavani (Sreenivasan and Sounderraj, 1967).

Parameters	Above the outfall	At the outfall	0.8 km below outfall
pH	7.3-8.8	4.6-11.0	6.5-9.0
Conductivity (μ mhos)	22.7-30.0	26.4-36.5	24.0-31.9
Free mineral acid (mg/l)	Nil	0.0-280.0	Nil
Hydroxide (mg/1)	Nil	0.0-240.0	Nil
Dissolved oxygen (mg/l)	4.0-9.4	0.0-7.8	2.0-8.6
Total solids (mg/l)	62-148	1688-8744	138-240
Suspehded solids (mg/l)	16-96	150-412	48-130
BOD (mg/l)	5.4-6.2	55-120.4	–

River Gadavari receives about 45,000 gallons/day black liquer from Andhra Paper Mills and due to this population is localised in the vicinity in a shallow zone of the river upto 1.5 km below the outfall. The high amount of putriscible organic matter reduces the oxygen of the river water upto 1.19 mg/l and the flocks of ammonium hydroxide covers surface and bottom of the river.

River Chaliyar at Calicut receives about 49641.7 m^3/day of Rayon factory effluents at Mavoor which have created a pollution hazard in this river, acutely in summer season. High BOD has been observed even upto a distance of 25 km down stream from waste discharge.

Pariyar river has become a dumping ground for all kinds of toxic effluents of the biological, chemical and industrial varieties. Liquid ammonia, ammonium sulphate, a variety of acids, mereury, zinc, sodium fluoride, DDT, BHC, radioactive materials like meso-thorium and uranium constitute the major pollutants. The high rate of the pollution of the river is posing a great threat to about three million people living in the Greater Cochin area, who depend entirely on this river for their drinking water requirements. Hundreds of tonnes of meso-thorium and uranium wastes being dumped into the sea and into the Periyar river reaches the human beings through fish, posing great danger of radiation induced cancer.

Enormous use of water in the condensors of Thermal Power stations leads to changes in temperature and physico-chemical properties causing deleterious effects on the aquatic life. Fendar and Kharat (1992) studied toxic metal ions in waste water of 1100 MW Koradi Thermal Power Station near Bagpur (Table 63). The concentration of trace metals Copper, Lead, Zinc, Nickel, Cobalt, Cadmium and Manganese in the upstream and waste waters of the power plant were in the order of machine washing water > cooling effluent water > downstream water > upstream water. Further, the concentration of metal ions in downstream waste water was higher than in upstream water. Similarly, although the concentration of metal ions in machine washing water was higher than in cooling effluents water, the concentration of Cobalt was less in machine washing water. The higher concentration of metal ions in the upstream and waste water samples of Thermal Power station could be attributed to heavy contamination due to flyash.

Table 63 . Concentration of trace metals in fresh and waste water of Koradi Thermal Power station and also water collected from unexposed area in μg/ml (Fendar and Kharat, 1992).

Site	Cu	Pb	Zn	Ni	Co	Cd	Mn
Upstream water	10.2-12.2	0.85-1.30	41.0-45.0	3.5-5.8	6.2-6.8	0.20-0.45	42.3-502
Cooling erfluent water	15.0-30.0	21.75-25.25	74.0-85.0	4.75-6.25	16.75-20.25	1.2-1.5	4035-44.85
Machine washing water	45.0-75.0	18.0-25.0	188.0-208.0	7.0-10.0	4.8-7.0	1.25-135	52.0-68.0
Downstream water	13.8-15.7	1.43-1.82	61.5-64.0	63-7.8	6.9-7.8	0.25-0.72	55.0-67.0
Water collected from unexposed area	7.3-8.25	0.55-0.66	32.0-33.2	2.5-2.75	5.2-5.5	0.15-0.20	31.0-31.5

Kalu river in Maharastra is contaminated by several industries, chiefly chemical and textile (Sharma, 1977). The last part of Kalu near Kalyan is so polluted that its pH has been recorded as low as 1.2 to 2.4 denoting sheer acid conditions and thus form an acidic barrier. Similarly, Krishna river in Maharastra is subjected to several diversions for irrigation, industrial and domestic use. Their wastes are dumped every day threatening the quality of river water. Boralkar, et.al., (1981) reported that there was an increase in the salt build-up at an alarming rate in this river, as a result several hundreds hectares of land have already become worthless.

River Daha receives about 45,000 gallons/day of putriscible organic waste, having 8090 to 8750 mg/l BOD, 16,693 to 18,644 mg/l total solids and 5.3 pH is discharged by four Sugar mills, one Distillary and one paper factory (Thakur and Deshpande, 1976). The total BOD load discharged in this river worked out approximately 12,192 m tonnes/day, which causes year round pollution and anaerobic conditions (oxygen level varying between 0.0 to 1.5 mg/1 in the 44 km stretch) in the river except during rainy season. This waste with low pH and high oxygen demand creates a chain of reduction and oxidation in the river involving iron compounds.

River Damodar, which is the main source of water both for domestic and industrial use, in the higly industrialised tract of Dargapur-Asansol in West Bengal, is suffering from serious pollution. A large number of industries such as Sindri Fertilizer plant, the Bihar government superphosphate factory and sevral associated cement units are located on the bank of this river. The daily discharge of 18,000 m^3/day of wastes contain alkalies, chromates, ammonia, cyanides, phenols, nepthaline etc. Having about 43,000 kg BOD. Along with this three Thermal Power Plants (Patratu, Bokaro, Chandrapur), Kargoli coal washery discharge finally suspended soal particles into this river. The Bokaro thermal power plant also releases fly ash and Gargali coal washery about 40.04 mt/day of fine metallurgical coal into this river, thus deteriorating the water quality. It has been estimated that 600 ton/day ash and or coal particles are "distributed into this river by the Thermal power plants and Coal washeries (Thakur and Deshpande, 1976).

Hoogly river (a 100 km stretch) has been found to contain 350 outfalls on either side of the river pouring out large quantities of industrial effluents, adding approxinately 20 x 10^6 m^3/day wastes. Drinking water from Hoogly was hazardous to health as more than 150 factoreis including 87 jute mills and 12 textile mills drains untreated waste into the river.

During the last few years much concern has developed over the continued deterioration of lake system. This is mainly due to uncontrolled human activities which are responsible for the accelerated flow of materials from the terrestrial to the aquatic portion of the water shed. Fresh water bodies (Pichhola, Rang Sagar, Swaroop Sagar, Faten Sagar and Udai Sagar lakes) with their picturesque surroundings is a dominant feature of Udaipur city. All these lakes differ in respect of human interferences (Vyas, 1986).

Fateh sagar, basic irrigation reservoir is now supposed to serve in addition numerous other functions of touristic needs, domestic water supply and daily bathing and washing demands of growing city. The increasing number of establishments-touristic hotels and restaurants, parks, gardens, residential apartments and aggregation of hockers account for the problems of added pollution to the lake.

Pichhola lake caters the needs of Udaipur inhabitants for all types of domestic purposes. The main causes of pollution of water of this lake are washing and bathing, disposal of untreated sewage and washing in of agrochemicals.

Rangsagar lake is being polluted by washerman's activity and by the disposal of untreated sewage and solid waste by the inhabitants of the surrounding areas.

Swaroop sagar lake serves people of Udaipur in numerous ways particularly for daily domestic needs and by virtue of its neighbourhood location to city proves to be of immense utility and direct approach. The washerman's of the city use it day and night. The growing touristic demands, residential pressure and beautification programme have collectively utilized its locational aspect to their welfare but unfortunately at its own cost and peril. On account of the construction of new colonies, the vegetation of the hillocks around it has been removed and rocks are exposed. Due to this a large amount of silt in addition to domestic sewage from the residential colony fall into this lake. The most conspicuous feature threatening this lake is that the people living around this lake having no sanitation facilities, directly come to the lake and sit on the foot steps of the lake. As a result a large amount of human extreta gets accumulated and ultimately falls in the lake water.

The important source of pollution of Udaisagar lake are city sewage and effluents of various industries of the industrial area around Udaipur town. The most important of them worth mentioning are Udaipur distillary and Reliance Chemotax.

Important biochemical estimation of these lakes waters have been given in Table 64.

Table 64. Biochemical characteristics (ppm) of water of five lakes in and around Udaipur (Vyas, 1986).

Lakes	Chlorides	Nitrates	Phosphate	BOD
Rangsagar	54.98-189.9	0.20-1.22	0.20-1.40	7.6-14.2
Swaroopsagar	64.96-179.94	0.40-1.88	0.18-0.96	8.20-16.2
Fatehsagar	82.5-146.34	0.7-1.46	0.5-0.71	4.6-12.0
Pichhola	65.60-124.32	0.34-1.23	0.32-1.68	8.3-19.4
Udaisagar	104.96-288.0	0.36-1.23	0.32-1.68	8.3-19.4

Several workers have reported organic pollution to influence the chloride concentration. Sreenivasan (1964) is of the opinion that chloride concentration between 4 to 10 ppm indicates purity of water. Adoni (1975) has reported that a chloride concentration above 60 ppm indicates heavy pollution. The values of chloride concentration for Udaipur lakes are indicative of heavy pollution. Since untreated city sewage is thrown directly in these lakes the concentration of chlorides naturally increases.

The nitrate - nitrogen of these lakes have been severely altered by increase in loading of inorganic and organic nitrogenous compounds. Nitrates enter the aquatic body from various sources like erosion of natural soils and artificially fertilised soils through rainfall and sewage (Shekhwat, 1983). In the lakes studied highest concentration of nitrates was recorded in Fatehsagar and Swaroopsagar.

The phosphate levels for Udaipur lakes are indicative of high phosphorus level like that of nitrogen. The main natural origin of phosphate is due to chemical and mechanical weathering of rocks. The other most important source of phosphate is sewage and detergents. However, the leaching and drainage of fertilizers from the agricultural fields of the catchment area is another factor responsible for increasing phosphate concentration in these lakes.

Biochemical oxygen demand values in these lakes have been observed to be quite high, ranging from 4.6 ppm to 19.4 ppm. These high values of BOD are indicative of water enrichment specially due to pollution through kitchen refuse and sewage (Zutshi and Vass, 1973).

The status of various biochemical properties of water suggest that nutritional load is very high in these lentic bodies. Thus keeping the above comparative biochemical findings in view, it may be concluded that the water of Udaipur lakes is much stressed as well as polluted.

There are a number of areas in this country where serious ground water pollution problems have been caused due to the seepage of effluents from various sources. Olaniya and Saxena (1977) reported that the 25 wells around a refuse dumping site in Jaipur have been found to have high dissolved solids, chlorides, iron salts, COD and hardness. The effect was noticed upto a distance of 450 metres. Siva and Ramamoorthy (1977) reported that the well waters were generally high in mineral contents and had harder water in the areas where subsurface methods of sewage disposal were practiced. The chlorides and total dissolved solid level was very high in these wells. Narayan and Madhyastha (1985) reported that water of well in Mukka village near Mangalore have become polluted (Table 65) as oily effluents seeped continuously from the north-eastern and the south-eastern sides of the well at a distance of 10 metres, dumping pits of a fish oil extracting small scale industry.

Table 65. Physico-chemical parameters (mg/l) of water samples in Mukka, near Mangalore (Narayana and Madhyastha, 1985).

Source	Temp	pH	Total hardness	Total Solids	Alkali-nity	CO_2	DO	BOD	GOD	Oils
Seepage water	27.5	8.5	530	626	350	30	5.0	14	84	2 1 0
Polluted water	26.0	8.6	532	620	356	28	5.4	12	82	236.6
Control	25.0	7.6	260	340	24	6	7.8	2	8	nil

HEALTH HAZARDS OF WATER POLLUTION

Water, as a part of the human environment, occurs in four main forms - as ground water, fresh water on the surface, in the sea as salt water, and as vapour in the atmosphere. Human health may be affected by ingesting water directly or in food, by using it in personal hygiene or for agriculture, industry, recreation, or by living near it. In general polluted water may cause several types of diseases such as:

		Diseases
1.	Water borne	
	Water which has been contaminated by poor sanitation acts as vehicle for infecting agents.	Cholera, typhoid, infectious hepatitis,
2.	Insufficient available water to allow people to wash regularly, infections develop.	Scabies, yaws, leprosy, trachoma.
3.	Water based	
	Essential paet of life cycle of infecting agent takes place in aquatic animals, persons drink or walks in water.	Schistosomiasis, guinea worm.
4.	From water related vectors	
	Infection-caryying insects breed in water and bite near it specially when stagnant.	Malaria, sleeping sickness, yellow fever.

PORTS OF ENTRY OF TOXIC WATER POLLUTANTS

The toxic water pollutants can gain entry into the human system through the following two ports:

1. The oral route of ingestion with the gastro-intestinal tract as the primary target along with associated organs of absorption, digestion, biotransformation and excretion.

2. By penetration of the pollutants through general skin or mucous membranes, or by insect vectors bite.

HAZARDS ASSOCIATED WITH INGESTION OF POLLUTED WATER

The pathogens transmitted directly by water or indirectly through water to food, constitute one of the main sources of diseases. They include animate pathogens such as bacteria, virus, protozoans, worms, helminths and inanimate pathogens such as inorganic and organic chemicals and radioactivity.

Bacterial infection among rural and urban population is of common occurrence. Table 66 shows the principal diseases attributable to the infection of water borne bacteria.

Table 66. Human diseases through ingestion of animate pathogens in contaminated water.

Diseases	Causative organism
Cholera	*Vibrio cholera*
Bacillary dysentry	*Shigella spp*
Typhoid	*Salmonella typhi*
Paratyphoid	*Salmonella paratyphi* A,B, C.
Gastroenteritis	Other *Salmonella* types, *Shigella*, *Proteus* spp.
Infentile diorrhia	Enetropathogenic types of *Escherichia* coli
Leptospirosis	*Leptospira* spp
Tularaemia	*Pasteurella tularensis*
Infectious hepatitis	Virus not yet Identified
Intestinal amoabiasis	*Entanoeba histolytica*
Dranontiasis	Guinea worm
Distomatosis	*Fasciola* spp., *Dicrocoelium* spp.
Ascariosis	*Ascaris lumbricoides*

During 1960's, classical Cholera caused by *Vibrio cholerae* had receded in Calcutta. However, cholera "E1 Tor" which emerged in 1961 from its endemic foci in Indonesia, has spread in many countries in the western Pacific and in the south-east and central Asia. During 1970 series of outbreaks of Cholera "El Tor" occurred in areas not normally affected e.g., the eastern Mediterranean and the USSR, as well as a number of African countries. In 1971, cholera spread to nine more African countries, and small outbreaks or individual cases of cholera occurred in six European countries'. Typhoid and Paratyphoid fevers are still widely disseminated throughout the world (Bernard, 1965).

The viruses (not yet identified) of infectious hepatitis (Chang, 1968) has a global distribution, in the polluted waters (Koff, 1970). A striking example was the epidemic of infectious hepatitis in Delhi (1955 - 56), in which more than 28,000 cases were identified, with a case fatality of 0.9 per 1000 (Viswanathan, 1957).

Among the protozoans, *Entamoeba histolytica* is the causal agent of both intestinal amoebiasis (e.g., amoebic dysentry and its complications) and extra- intestinal forms of the disease, such as amoebic liver abscess. It is widespread throughout the tropical countries of the world and wherever sanitary conditions are poor. Amoebic cysts are resistant to chlorine in doses normally applied in water treatment. (WHO, 1969).

The guinea-worm, which causes dracontiasis is common among the rural populations of many developing countries. This parasite is transmitted through open village wells and ponds infected with the copepods intermediate host.

Some intestinal helminths, such as *Ascaris lumbricoides* and *Trichuris trichiura* may be water borne. Distomatosis is another parasitic disease that may be contracted by swallowing contaminated water containing cysts of *Fasciola* and *Dicrocoelium*.

High concentration of nitrate in surface waters result in infant methaemoglobinaemia (Schmidt and Knotek, 1970). Nitrate poisoning have been frequently noticed in various districts of Rajasthan and alleged to have caused infant deaths in the state.

If the flouride in drinking water is less than 0.5 mg/1, the incidence of dental caries in childern is likely to be high (McClure, 1970). However, when present in much greater concentration, they can cause endemic cumulative fluorisis with resulting skelatal damage (Table 67) (Adler, et.al., 1970).

Table 67. Diseases associated with inanimate pathogens.

Diseases	Chemicals	Concentration
Methamaoglobinaemia	Nitrates	20-300 mg/litre
Dental caries	Fluorides	less than 0.5 mg/litre
Fluorisis	-do-	6-12 mg/litre
Black foot	Arsenic	0.24-0.96 mg/litre
Minamata disease	Methyl mercury	1-10 ug/litre
Shortened life	Lead	0.05 mg/litre
Itai-Itai	Cadmium	1-10 ug/litre
Cardiovascular	Hardness water	

Rajasthan fluorisis problem has reached threatening proportions. It has permanently crippled over 3.5 lakh inhabitants and is likely to cripple many more. Banka patti (Nagaur) is among the several villages in the state where fluoride pollution poses serious health hazard. Rajasthan is believed to have the maximum number of humped back persons because of high fluoride concentration in water sources.

Fluoride first result in mottling of teeth and outward bending of legs from the knees. Prolonged intake of fluoride containing water stiffens the bone joints, particularly those of the spinal cord. The hump once developed can not be cured.

Water borne Arsenic causes "black foot" disease (Cheng Chi and Blackwell, 1968). It accumulates in some marine organisms, such as dams and shrimps.

The well known outbreak of *Minamata* disease in Japan is associated with methylmercury poisoning, where the people consumed polluted fish and shellfish (Tsuchiya, 1969). The Minamata disease, in short is a severe intoxication of Central nervous system, caused by Methyl mercury. It was identified as a disease in May 1956 by Dr.Hajime Hosokawa (1901-1970).

The following facts were established:

1. The first case of Minamata disease appeared at the end of 1953 and several cases occurred in the following years in the same district.

2. In 1956 the disease took epidemic form.
3. Observed symptoms were quite new, unknown once in the available literature.
4. Some what similar symptoms were recognised in cats in the region before human symptoms appeared, and generally those cats jumped into the sea.
5. The patients were mostly fishermen, or their families, and there was a tendency of successive occurrence of patients in the same family.
6. The patients were big consumers of Minamata Bay fish and they ate fresh fish as their main food.
7. The disease was not infectious.

Mercury as a chemical compound affects the central nervous system. It damages the brain and spinal cord so that motor co-ordination of all kinds is impaired, resulting in loss of the ability to walk, speak, write and feel. In the victims of this kind of poisoning, depending on the level of contamination higher human functions get ruined and only the very base of animal functions were left some what active. In the worst cases of mercury poisoning resulted in death of very agonising sort (Carter, 1973). The disease is on the march and spreading to other parts of Japan, perhaps it will spread to other parts of the world (Ui, 1971).

Itai-Itai is a specific disease, observed in Toyama city in Japan. It has been attributed to cadmium contamination of Jintsu river and surrounding rice paddies. It is known as Itai-Itai (Ouch-Ouch) because of the painful symptoms from multiple fractures arising from ostemalacia. It appears to be largely confined to post-menopausal women who have experienced several deliveries. Later, skeletal deformation take place, with marked decrease in body height, proteinuria and glaucoma. There is an increase in serum alkaline phosphate, and a decrease in inorganic phosphorus.

Hardness of drinking water is invariably correlated with death rate from cardiovascular diseases. Areas supplied with soft drinking water almost consistently experience a significant higher prevalence of either arteriosclerotic heart disease, or degenerative heart disease, hypertension, sudden deaths of cardiovascular origin, or a combination of these.

Excessive levels of DDT and PCB and other chlorinated hydrocarbons affect functioning of liver, could affect chlorestrol level and might impair the development and functioning of nervous system. DDT and other insecticides have already been branded as potential human carcinogen. DDT also inhibits the absorption of calcium by bones. DDT produces enzymes which decompose sex hormones steroids thus creating hormonal imbalances. The most important factor influencing the hazard from pesticide is dosage. There is no compound so safe that it cannot be dangerous and even fatal if sufficiently misused. Xintaras, et.al.(1979) listed sixteen symptoms of neuro-toxicity of Leptophos in men:

1. Drooping of eyelids, blurring of double vision.
2. Decrease in memory or thinking ability.
3. Difficulty in speaking.
4. Paralysis of any part of the body.
5. Loss of balance or staggering.
6. Numbness of tingling in hands or feet.
7. Difficulty in walking.
8. Problem with coordination.
9. Spells if dizziness.
10. Muscle weakness .
11. Nervous or uncontrolled tension.
12. Frequently feel fatigued.
13. Difficulty in sleeping.
14. Change in handwriting.
15. Drowzy or sleeply during the day.
16. Unexpected sweating.

Many polynuclear aromatic hydrocarbons (PAH) in the concentration of 0.001 to 0.100 μg/litre found in water are known to be carcinogenic (Borneff and Kunte, 1964).

The radioactivity of water from natural areas is usually low and of no immediate health significance. Pollution by radioactive wastes however may be dangerous. The most common radionuclides in drinking water are ^{226}Ra, ^{222}Rn, ^{232}Th and ^{238}U. They are also known to be carcinogenic. Fallout from nuclear testing, discharge from nuclear power plants and reprocessing plants and the disposal of radioactive wastes release ^{90}Sr, ^{137}Cs, and ^{131}I which are of importance in relation to health hazards. They may be ingested directly through water supplies, or may be concentrated in fish and edible seeweeds and cause hazards.

The radioactive elements enter into human bodies either directly or indirectly. There it accumulates to produce leucemia, bone cancer and heredity diseases. The cumulative increase of radioactive elements in the body can be calamitous to the human race.

HAZARDS ASSOCIATED WITH POLLUTED WATER CONTACT

Direct contact with water, occurs mostly in recreational activities - swimming, water skiing, washing of clothes, disposal of human excreta, industrial effluents discharge etc. so that these waters become highly polluted. The animate and inanimate pathogens of such polluted waters penetrate through skin or mucous membranes and cause health hazards.

Schistosomiasis is a chronic, insidious, delibitating disease that may cause serious pathological lesions, saps energy, lowers resistance and reduces output of work. It is chiefly due to three species of trematodes, namely *Schistosoma mansoni*, *S. japonicum* and *S. haematobium*. Of these the first two give rise to intestinal manifestations and the third to genito-urinary or vesical schistosomiasis. The eggs, released is faeces or urine, hatch as miracidia in water, where they penetrate snails. They emerge from the snails in the form of cercariae ready to infect man. Penetration is through skin, while wading, bathing etc.

In many parts of the world, bathers in lakes may be affected by "Swimmer's itch". This dermatitis is caused by the penetration of the skin by cercariae of schistosomes from birds and rodents (Table 68).

Table 68. Human diseases capable of being transmitted through water contact other than ingestion.

Diseases	Cause	Agent
Internal Schistosomiasis	*Schistosoma mansoni* *S.japonicum*	Skin penetartion of snails carcariae while bathing.
Genitourinary schistosomiasis	*S.haemetobium*	
Swimmer's itch	-do-	Skin penetration of birds and rodents carcariae.
Leptospirosis	*Leptospires spp.*	Skin and mucous membrane penetration.

Leptospirosis is a bacterial infection in the natural hosts (wild animals and domestic animals) that excrete leptospires in urine and man is infected through the skin and mucous membranes from contact with the water of contaminated ponds, canals, rivers etc.

Synthetic detergents are used in the manufacture of washing powders, shampoos and tooth pastes. Sodium alkyl sulphate is synthetic surfactant commonly included in dish detergents and shampoos. Biodegradable LAS is some respects is more hazardous than branched ABS. The most effective and used most widely is Sodium tripolyphosphate (STP). Residues of alkyl benzene sulphate (ABSO) used as surface active components in sythetic detergents mixtures which penetrate the skin, spread through the body and into the womb. Thus it is undoubtedly dangerous for the developing child of the pregnant women. Linear alkyl benzene sulphonate (LAS) is a linear form of ABS molecule. Because of its branching structure the original ABS molecule is highly resistant to biological decomposition.

Synthetic detergents remaining on our hands after washing dishes does not loose its property of working to break down oil, and therefore attacks the natural, protective thin layer of oil on our skin. The result is dry rough and even badly cracked skin. The damage to infants skin is more serious. The synthetic detergents cause numerous bald or thin haired people. They contribute birth defects via pregnant women. ABS may cause damage to the offspring via male sperms. Even DNA structure had been altered, which could create physically deformed or genetically imperfect future generations.

There are also other dangerous attributes of synthetic detergents, such as its ready permeation through human skin. Mercury and other heavy metals and PCB when brought in contact with synthetic detergents, have greater efftects and chances of entering our bodies. This is one of the most grave aspects of the detergent problem and a cause for considerable concern (Ui, 1971).

HAZARDS ASSOCIATED WITH WATER ASSOCIATED INSECT VECTORS

Of the diseases associated with water associated insect vectors, the most widespread is Malaria, transmitted by the female Anopheles mosquitos.

River blindness or Onchocerciasis which occurs near clear springs over rocky beds, where a large proportion of the resident population may become partially or completely blind, is transmitted by balck flies.

Yellow fever or sleeping sickness, which is widespread in Africa is transmitted by the tsetse fly.

Filariasis which affects more than 250 million people throughout the world, is transmitted by the major vector Culex pipiens fatigans (Table 69).

Table 69. Diseases capable of being transmitted through water associated insect vectors.

Disease	Causative organism	Insect vector
Malaria	*Plasmodium spp*	Anopheles mosquitos
Riverblindness	*Onchocera volvulus*	Simulium spp (blackflies)
Yellow fever	*Arbovirus* group B	Aedes mosquitos
Sleepingsickness	*Tryjpnosoma gambiense*	Tsetse fly
Filariasis	*Wuchereria bancrofti*	Cules pipens fatigans
	Brugia malayi.	mosquitos.

Chapter-3

THERMAL POLLUTION

Thermal pollution can be defined as an accumulation of unusable heat from human activities that disrupts ecosystems in the natural environment. The most important anthropogenic source of thermal pollution are the industries which reject vast quantities of heat in the environment. Vast quantities of water find use for cooling purposes by steam electric power generation stations (and other industries to a lesser extent). Cooling water is discharged at a raised temperature, at 8-10 °C higher than the coolant uptake water. As a result, close to the outfall, there exists an intermixing zone with water temperatures slightly higher than the ambient waters.

Increasing the water temperature of a system is harmful since it generally alters the physical, chemical and biological characteristics of that system. Aquatic systems are most delicately balanced ecosystems which do not fluctuate much in temperature as do the land masses. Therefore, a slight change in temperature over the natural limit will affect them to a great extent. The increase in heat contributes to the physical, chemical and biological changes in the receiving waters. These changes can be beneficial, detrimental or insignificant, depending upon the ecology of the receiving water body, the desired uses of that water body, and the amount of heat discharged. The bahaviour and impact of pollutants in water also modified at elevated temperatures (Table 70).

Table 70 . Behaviour and impact of water pollutants or elevated temperatures.

Impact/Behaviour	Qualitative magnitude
Solubility of gases	low
Solubility of liquids and solids	high
Biological uptake rates	high
Biological release rates	high
Rates of physico-chemical degradation	high
Rate of biological degradation	high
Toxicity of pollutants	high
Toxicity thresholds	low
Rates of chemical oxygen depletion	high
Rates of biological oxygen depletion	high
Biological impact of nutrients	high
Biological impact of suspended solids	high
Rate of denitrification	high

Heat is not ordinarily thought of as a pollutant by many people, atleast not in the same sense as a corrosive chemical. However, the addition of excess heat to a body of water brings about adverse effects as numerous as many of the chemical pollutants. This serious problem of thermal pollution originates primarily with the practice of using water as a coolant in many industrial processes. Most water used for this purpose gets returned, with the added heat, to the original sources.

Power generated by coal powered or nuclear powered stations require cooling water for heat removal. Other industries like textiles and sugar also release heat but to a much lesser extent. A large proportion of the electricity used is produced in turbogenerators using steam power. To increase turbine efficiency, a partial vacuum is created at the turbine exhaust by cooling and condensation of the turbine steam. The extent of cooling and therefore, of vaccuum achieved in practice is limited by consideration of heat transfer. In modern power stations producing 100 MW, nearly one million gallons of water are discharged in an hour with increase in temperature of the cooling water passing by 6 to 9 °C.

Of the power stations installed, about 40 percent use estuarine or sea water for cooling purposes. In these, the cooling water is pumped direct to the condensers and returns warmed as far as possible from the point of intake. This is known as the "direct" or "open" system of cooling. The heat output is greater because of increased load, during day, as compared to night.

The efficiency (heat power) of fossil fueled power station is about 40 percent and of boiling water reactors in operation about 33 percent. Loss in efficiency and production of "waste heat" is due to in-plant and stack losses. Such losses are taken as 15 percent of the plant heat rate. Thus heat of cooling water (BTU/kWh) 0.85 x heat rate - 3413, Nuclear heat to cooling water = 0.95 (10-342)- 3413=6400 BTU/kWh (3413 BTU) is required for 1 kWh at full efficiency. The quantity of heat rejected to cooling water from nuclear power stations may be computed in the same manner. Typical in plant losses are about 5 percent of the above equation is rewritten as:

Heat to cooling water (BTU/kWh) = 0.95 x heat rate-3413

As an example, the quantities of waste heat released in the two types of power stations are given below:

For fossil fuel at 40 percent efficiency

$$\text{Heat Rate} = \frac{3413}{n1 \div 100} = \frac{3413}{0.40} = 8533\,\text{BTU / kWH}$$

Heat to cooling water = 0.85 (8533)-3413 - 3800 BTU/kWh

For nuclear fueled power stations at 33 percent efficiency

$$\text{Heat rate} = \frac{3413}{0.33} = 10.342\ \text{BTU/kWh}$$

Calculated as above, a 100 MW electrical output nuclear power station with 33 percent efficiency will release to cooling water 6.41 x 10^9 BTU/h waste heat from a single power station at 100 MW will riase the temperature of 3000 cfs flow of river by 10°F (6°C).

EFFECTS OF THERMAL POLLUTION

Thermal pollution exerts physical, chemical and biological affects:

Physical Effects

The temperature variations effect almost every physical property of water. Temperature influences the viscosity, density, vapour pressure, surface tension, gas solubility and gas diffusion rates (Table 71). Heated water has low density and spreads on the surface of water, causing them to stratify thermally. The stratification is a barrier to the oxygen penetration into the deeper layers. This also disrupts the normal circulation patterns, the ecological consequences of which would be drastic, unpridictable and almost certainly deletewrious. At elevated temperatures, the sedimentation of suspended materials increases due to reduction in density and viscosity of water. Evaporation rate of water increases at high temperature. Warm water reduces its palatability. Once the receiving water becomes warm, it is not suitable further as cooling water because of the decrease in efficiency of heat transfer.

Table 71. Physical properties of water at various temperatures (After Parker and Krenkel, 1969).

Temperature	Vapour pressure (mm Hg)	Viscosity (centipoise)	Density (g/ml)	Surface tention dynes/cm	Oxygen solubility (mg/L)	Oxygen diffusivity (cm^2xsec x 10^{-6})
0	4.579	1.787	0.99984	75.6	14.6	–
5	6.543	1.519	0.99997	74.9	12.8	–
10	9.209	1.307	0.99970	74.2	11.3	15.7
15	12.788	1.139	0.99910	73.5	10.2	18.3
20	17.535	1.002	0.99820	72.8	9.2	20.9
25	23.756	0.890	0.99704	72.0	8.4	23.7
30	31.824	0.798	0.99565	71.2	7.6	27.4
35	42.175	0.719	0.99406	–	7.1	–
40	55.324	0.653	0.99224	99.6	6.6	–

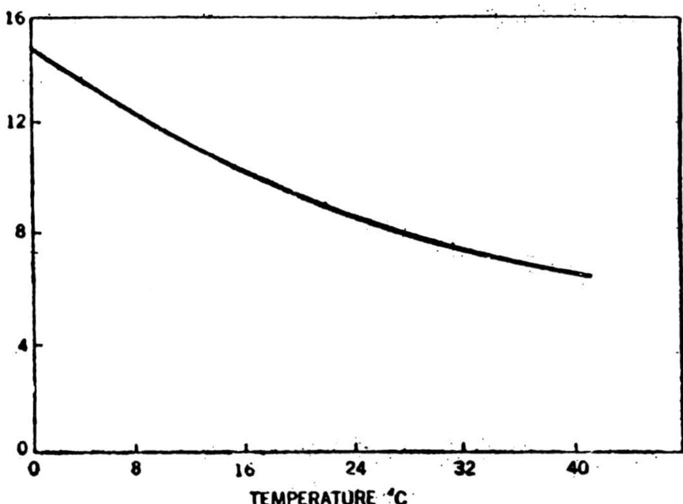

Fig. 17. The quantity of oxygen dissolved in water, is related to temperature. The higher the temperature, the lower the oxygen content of the water. Thermal pollution can lower the oxygen content below the point necessary to sustain many animals.

Chemical Effects

Chemistry of waters greatly depends upon the temperature. Rate of chemical reactions normally increases with rise in temperature which is about two-fold with rise of every 10°C. The decreasing ability of water to contain dissolved oxygen as the temperature increases have been shown graphically in Figure 17.

The addition of heated water to a cooler body of water may accelerate the lowering of dissolved oxygen levels due to density differences between the too. The less dense warm water tend to form a layer on top of the cooler, more dense water. This occurs particularly when the body of cool receiving water is deep. The resulting blanket of "hot" water cannot dissolve as much atmospheric oxygen as the underlying cold water, which is denied contact with the atmosphere. Normal biological reduction of the dissolved oxygen level of the atmospherically unreplanished lower layer may give rise to anaerobic conditions.

Another effect of this stratification may show up downstream from a dam when the oxygen deficient lower level get discharged through the lower gates of a dam. Serious effects on downstream fish life may result. Also, the ability of the stream below the dam to assimilate oxygen-demanding wastes will get curtailed.

Biological Effects

The biological effects of thermal pollution include effects on metabolic rates, reproductive rates, longivity, acclimation etc.

(a) Metabolic Rates

Fishes show a marked rise in basal rate of metabolism with temperature to the lethal point. In brown trout, the rate of oxygen consumption rose steadily till lethal temperature (27°C) was reached. The uptake of food, respiratory rate and swimming speed increases with rise in temperature. Sockeye salmon cruises twice as fast in water at 15°C as at 0°C, but above 15 °C its speed declines.

At hish temperature, the dissolved oxygen decreases, while the metabolic rates of the organisms, requiring oxygen, increases, thus accebtuating the stress. At the same time the bacterial activity increases, further reducing oxygen supply. The water may rapidly become unfit for all but for a few anaerobic species. High temperature works as barrier for oxygen penetration into cooler deep waters. The aerobic degradation gives way for the anaerobic degradation, making the water more polluted. Further, in organically polluted waters, multiplication rate of bacteria increases with increase in temperature, especially where the food supply is in plenty.

(b) Reproductive Rate

The rising temperature triggers deposition of eggs by female. The triggering is particularly dramatic in estuarine shellfish (Oysters and clams), which spawn four hours of the water temperature reaching the critical level. The Atlantic salmon eggs hatch in 114 days in winter (2°C) and in 90 days at 7°C. Harring eggs hatch in 47 days at 0 °C and in 8 days at 14.5 °C. Trout egg hatch in 165 days when incubated at 37°F. However, at water temperature of 54°F only 32 days are needed and no hatching takes place at water temperature in excess of 59°F. Such a result can be disastrous to fish population. These are critical temperatures for reproduction. At temperature 22 °C and above the banded sunfish fails to develop eggs. Temperatures in the range of 20-23°C prevent cell division in carp. Gammarus above 7.8°C produces only female offspring.

(c) Longivity

High temperature may induce increase in activity, which exhausts the organism and shortens life. Crustacean *Daphnia* lives for 108 days at 7.8°C and only 29 days at 21°C. In assessing the total hazard present, temperature effects on biochemical systems, degradation and catalytic influence should be considered. In general, the speed of a chemical change is approximately doubled for each 10°C rise in temperature. The majority of organisms affecting chemical water quality are in the mesophytic classification and thrive in a temperature range of 10-40°C. For this group, activity usually reaches a maximum between 30-37°C, then falls off as enzymes become less active.

When the temperature of water containing biodegradable matter rises, the intensified action of micro-organisms causes the biological oxygen demand to be accomplished at a lower temperature. Taste and odour problems induced by temperature accelerated chemical or biochemical action or accentuated, when oxygen is depleted. The synergistic effect may have to be considered in assessing the hazard due to multiplicity of polluting factors.

Synergism may be defined as the accumulating result of the action of separate agents, which together have a greater impact than the sum of their individual contributions. At 10°C rise in water temperature doubles the toxic effect of Potassium cyanide and an 80°C rise tripples the toxic effect of O-xylene. Since domestic and industrial processes waste are always present in the effluents, the synergistc action in water is always present in effluents and is relatively a common phenomenon. The effect of rise in temperature leading to lowered dissolved oxygen content in such cases increase toxicity of poisons present.

(d) Acclimation

Faced with continued rise of temparature from discharge of hot effluents, fishes adapt to the new environment by adjusting the body processes. This is particularly so in the case of benthic and sedentary creatures, who live close to the shore. These non-genetic adaptations can be described in three stages:

(i) Immediate response (like shock)
(ii) Stabilisation process (gradual adjustment)
(iii) New steady state.

The immediate response of rising water temperature in a water body receiving heated collant water is the quick death of fishes. The thermal death may be due to the action of heat on the nervous system, coagulation of cell protoplasm or the inactivation of enzymes.

The stabilization process varies with species. In *Homarus americanus,* thermal acclimation (14.5 - 23 °C) was complete in three weeks. In some fishes, it may take only four-five days. The capacity to acclimate to higher temperature depends on the genetic background, environmental history, physiological conditions and age of the fishes. The rate of acclimation depends upon metabolic rate. The acclimation process is rather gradual and limited unlike in warm-blooded animals. Any change in temperature is immediately communicated to the blood in the gills, and the blood passes them to the whole body of the fish. The fish have no system of heat compensation except moving away from the torrid waters. In addition, the haemoglobin of fish blood has a reduced affinity for oxygen. The carp can exist at 0.6 mg/1 of oxygen at 0.5°C and needs 1.5 mg/l at 35°C. The metabolic activity increases with temperature and tapers off as lethal temperature are reached. *Astacus,* a Crustacean showed 30 heart beats per minute at 4°C, 125 at 22°C and 65 at 35°C, which is approaching the lethal temperature.

(e) Temperature lethal to fish

If temperature rises rapidly, fishes die quickly depending upon the temperature and acclimation temperature. The sensitivity of fish to temperature can be obtained by substracting the acclimation temperature from the lethal values The possible effects of thermal pollution, therefore should consider three aspects. How does it effect dissolved oxygen ? What is the temperature range for heat death ? What are the pollutants present, which increase toxicity ?

The thermal death may be due to the action of heat on nervous system, coagulation of cell protoplasm or the inactivation of enzymes. They decrease as the acclimation temperature rises and indicate a level beyond which fish cannot be acclimatised.

Experiments have shown that mostly, fishes can get acclimated to a new condition. The tolerance temperature (T) have been calculated using the following equation:

$$T = \frac{tP_1 + (t-2)P_2 + (t-4)P_3}{100}$$

where, t is the highest temperature survived for the given period by the individual, (t-2)and (t-4) are temperatures 2 and 4°C below and P_1,P_2 and P_3 are survived at corresponding temperatures. Obviously, at any acclimation temperature, there are temperature zones, where existence is possible and zones, which are having a thermal death point. It is not enough if the fish kills are protected. It is important not to disturb the life-cycle.

Another factor to be considered in addition to the temperature change has been the rapidity with which the change occurs. Fishes are also able to acclimate themselves to the moderate temperature changes (below lethal levels), if the change has been not a sudden one. Normally a moderate temperature change over a 30-40 hour period could be handled by the fish. For instance, 95 percent of the eggs of Largemouth bass perish when suddenly transferred from water at 65-70°F into water at 85°F. However, if similar eggs are kept in water and the temperature gradually increased to 85°F over a 30-40 hour time period, 80 percent of the eggs survive.

(f) Preferred temperatures

When conditions are available, fish choose a final preferred temperature. Some trout show a maximum activity at a lower temperature. The lake trout has an upper lethal temperature of about 24 °C, while its maximum swimming speed is displayed at 15.5 °C. It has been shown that fishes can and will follow a gradient. They normally follow this gradient towards their preferrred temperature. Fish seek out the temperature that is best suited for survival. At preferred temperature (Table 72) fish congregate under natural conditions.

Table 72. Preferred Tempreratures for Fishes as Determined from Laboratory Experiments

Species	Degrees centigrade
Trout	13,16
Bluegill	32.2
Largemouth bass	30.0-32.2
Carp	32.0
Channel catfish	30.0-31
Pumpkinseed	28.1
Small mouth bass	28.0
Yellow perch	24.2
Yellow perch	21.0
Muskellunge	24.0

As different species favour different temperatures, thermal pollution may lead to the population decline of one species and growth of the another species. This results in shift of flora and fauna of the water body. For example, a cool water fish, trout, is frequently replaced under thermal pollution by catfish or others. At high temperature for relatively extended periods, the diatoms are first replaced by green algae and then by blue-green algae. In many cases temporary algal blooms have also been found to occur.

Chapter-4

INORGANIC POLLUTION

Water pollution due to the presence of inorganic pollutants is called inorganic pollution. Inorganic pollutants include inorganic salts, mineral acids, finely divided metals or metal compounds. These substances enter natural waters because of the activities in various smelting, metallurgical, and chemical industries; mine drainage and various natural processes.These pollutants bring about three general effects; the acidity, salinity, and toxicity of the water.

The inorganic chemicals of many types enter water from municipal and industrial waste waters and urban run-off. These pollutants are able to kill or injure fish and other aquatic organisms and they can interfere with the suitability of water for drinking or industrial use. Many of these inorganic pollutants are not only toxic but tend to concentrate in food-chains.

A prominant example is the occurrence of mercury in water. A number of industrial processes make use of mercury, some of which is eventually disposed of in effluents. The anairobic bacteria in the sediment convert inorganic mercury into Methyl mercury (CH_3Hg^+), which can be concentrated in aquatic organisms and lead to mercury poisoning.

One potential pollutant arises in petroleum drilling, where brine gets discharged along with crude oil when the latter is pumped out to the surface. In some places the brine have proved valuable source of important minerals and elements, such as Bromine, Iodine and Magnesium.

The acid mine drainage comes out from exposed coal mine surfaces, minerals, containing sulphur (most notably Iron pyrites FeS_2) in contact with air and water, forming Sulphuric acid that gets carried into streams by water draining from the mines. This takes place from abandoned mines as well as from operating mines and has been most pronounced in bituminous coal mines. Acid mine drainage has been the primary source of pollutants that increase water acidity. This discharge adversely effects thousands of kilometers of streams and has been one of the most significant causes of water quality degradation in coal producing areas. The actual pollutant present in mine drainage have been Sulphuric Acid (H_2SO_4) and soluble compounds of Iron. These substances are formed because of a reaction between air, water, and Iron pyrite, present in coal seams. Certain bacteria have been involved in the reaction, but their role has been not completely understood. This reaction can occur in both underground and surface mines.

During mining operations, in deep mines, the strata between the coal seam and the surface gets invariably disturbed. Fissures appear through which water drains into the mine from many surface areas. This water containing the harmful pollutants is invariably discharged into surface streams, either naturally or through man-made processes.

The presence of carbonates and bicarbonates in solution, maintains the pH of most productive freshwater between 6.5 and 8.5. This natural buffer system makes it possible for small amounts of acid or base to dissolve in the water without making appreciable changes in the pH. These ions also represent an indispensable reservoir for Carbon needed for aquatic plants during photosynthesis. Such use of these ions by aquatic plants can influence the buffering capacity of water, as there is a decided limit on the rate at which Carbon dioxide can be obtained from the atmosphere to replace that used by plants.

Large amount of salinity in water cause problems other than those related to human consumption. Dissolved inorganic and mineral substances exert adverse effects on aquatic animals and plants, and cause many irrigation problems. Damage to aquatic organisms is primarily related to the osmosis process, assuming the dissolved substances are non-toxic. Generally the concentration of dissolved materials in body fluids is the maximum that an aquatic organism can tolerate. When these organisms are in contact with water containing higher concentrations, there is a tendency for water to move out of the cells of the organisms into the surrounding water. The resulting increase in concentration within the cells of the organism can lead to death. Many freshwater species disappear when waters become brakish.

One of the most serious long-term effect of increased salinity of waters involves the use of and reuse of water in irrigation. It has been estimated that about 25 percent of the irrigated land of the world is affected to some extent by water salinity. Itrigation water brought onto a field always contains some dissolved salts. The concentration of these salts is in the range of 25-8000 mg/l. Plants extract water from the irrigated fields, but most of the dissolved salts are excluded by the roots. Water that evaporates from the soil surface leaves dissolved salts behind. These two processes make residual salts to accumulate in the soil.

The toxic properties of numerous inorganic compounds, especially those of some of the heavier metallic elements, have been well known. Some of these compounds are having desirable properties and so are routinely manufactured. Appropriate precautions are taken to insure the safety of individuals involved in the manufacturing process. The use of these compounds give rise to their introduction into the environment either directly or indirectly, intentionally or unintentionally. The detection of these metallic substances in air and water in concentrations approaching toxic levels has created a great amount of concern. The most toxic, persistent and abundant of these compounds in the environment appear to be those of the metals: Mercury, Lead, Arsenic, Cadmium, Chromium and Nickel. These accumulate in the bodies of organisms, remain for long periods of time and behave as *cumulative poisons.*

(a) Suspended solids

All bodies of water contain some quantities of suspended solids that may have originated naturally, or has been introduced as a result of human activities. The particles remain suspended for varying periods depending upon their size and density. The suspended matter may be either organic or mineral in nature. The organic suspended particles consists mostly of volatile solids derived from the detritus that creates an oxygen demand in water. A small portion of organic matter may also consist of synthetic non-biodegradable compounds which are sparingly soluble in water. The mineral suspended matter is mostly of natural origin such as silt with natural run-off, but some quantities can also be introduced by industrial activities.

The load of suspended solids, especially mineral matter, can vary from small quantities to thousands per litre in the natural water, depending upon the nature of soil, rainfall characteristics, activities in the catchment area and the nature of bottom sediments. The conditions usually leading to the excessive soil erosion like removal of vegetation, unmanned agriculture, and sandy nature of soil, may increase the silt load of run-off waters. During the rainy season, almost all bodies of water are heavily loaded with mineral particles.

The fate of suspended matter is governed largely by a number of factors operating in aquatic systems. The particulate matter in water do not remain suspended for indefinite periods, but tend to settle down with different settling velocities. The settling velocities of particles can be mathematically evaluated by the Stoke's relation as under:

$$\text{Settling velocity} = \frac{g.d^2p}{18\mu}(\rho_\rho - \rho)$$

Where,

d_p = diameter of the particle
m = viscosity of the medium
p_p = density of the particle
p = density of the medium
g = gravity constant

This indicates that finer particles are easily entrained remain in suspension for highly extended periods, whereas the larger and heavier particles deposit quickly.

The particles deposited at the bottom remain quite loose, and can be resuspended in water by the factors like convection, and diffusion both turbulent and brownian. However, the brownian diffusion is negligible in comparison to turbulent diffusion caused by wave action. Vertical convection is important, especially in areas of upwelling, and at the time of overturns.

The most important effect of the suspended matter in aquatic systems is the cutting of light by increasing turbidity. The penetration of light at any depth is the function of concentration of suspended particles and the extinction coefficient that can be expressed mathematically as follows:

$$l = l_{0-e}^{-kCL}$$

Where,

l_0 = intensity of light just below the water surface
l = intensity of light at depth L
k = extinction coefficient
C = concentration of suspended solids.

The value of extinction coefficient can be derived experimentally, but sometimes the values of Secchi disc light penetration can also be employed to evaluate extinction coefficient which remains almost equal to 1.9 divided by Secchi disc transparency (Golterman et al, 1978).

The reduction of light in the deeper layers promotes thermal stratification more readily. This may influence the biotic communities in deeper layers because of the depletion of oxygen in absence of its replanishment from the overlying layers.

The particles in suspension and those present in the sediments absorb cations, anions, and organic compounds, including several toxic materials like heavy metals, pesticides and PCBs. The fine grained particles, by virtue of their large surface to mass ratios, have comparatively higher adsorptive capacities than the larger particles. The suspended particles when settle onto the plants and animals, and dwellings at the bottom may cause high mortalities. Aquatic plants may suffer from mechanical damage and abrasion by settling particles (Alabaster and Lloyd, 1980). Larger invertebrates are affected by clogging of their gills and other respiratory surfaces. Besides, suspended particles can also clog other body surfaces of the animals. The suspended particles when deposited on the eggs of the organisms may effect them by impending the water and oxygen flow.

Turbidity in waters reduces the activity of fish and mat affect their migratory paths. Much lower quantities of suspended solids at the level of 35 mg/l can inhibit feeding in fishes. Suspended matter in general, reduces the diversity of life in aquatic systems. While the suspended matter of toxic nature can affect the organisms directly, the organic particles deplete the oxygen affecting the organisms indirectly. Suspended matter load, especially during rainy season, promote the silting of river beds which, in long run, may result in flooding of the area.

(b) Dissolved solids and hardness

The natural water never occur in the form of "pure water", but always have a variable quantity of dissolved inorganic salts. The commonly occurring natural salts are comprised mostly of cations like Sodium, Potassium, Calcium and Magnesium associated with anions like chloride, sulphate, bicarbonate and carbonate. They differ from the other pollutants in that they originate chiefly from the natural sources rather than mon-made sources.

The composition of solids present in natural body of water mainly depends upon the nature of the bed-rocks and the soil developed from it. The physico-chemical factors, which govern the chemistry of salts in water, may also influence the salt composition.

The salt content of the bodies of water often increases by the developmental activities in the catchment area. Large quantities of salts are picked-up from the soil by run-off to reach rivers and lakes. The drainage of irrigation water also leaches out significant quantities of salts from the soil that finally return to river as a highly saline water, or can contaminate the ground water.

The high salt content of the water bodies increases the soil salinity in several parts of the world, especially the arid habitats (Holmes and Talsma, 1981). The dams and reservoirs raise the ground water level in the surrounding areas and make the soil waterlogged. Continuous evaporation of this water from adjacent areas leads to an increased level of salts in the soil. Secondly the high salt content of the ground waters can charge the rivers during lean seasons. Thus, increasing their salinity, which may be harmful to the crops in case the water is used for irrigation. The waters exceeding 500 mg/l of dissolved salt and 750 μS of conductivity values are generally considered unsuitable for irrigation.

The high salt content in water is also not suitable for domestic water supply. At higher concentration exceeding 500 mg/l, the salts give a typical taste to the waters and reduce their portability. High concentration of salts near 3000 mg/l have been found to produce distress in cattle and livestock. For industrial use waters, the high salt content may produce scaling in boilers, corrosion of machinery and result in degraded quality of the product.

Theions, especially Ca^{++}, Mg^{++}, Cl^{-}, SO_4^{--} ,CO_4^{--} and HCO^{-}_3 impart hardness to waters. Hardness is the property of water which hampers the lather formation with soap. Hardness can be divided into temporary and permanent depending upon whether it can be removed by boiling the water. The salt of bicarbonate and carbonate produce temporary hardness, since they can be precipitated simply by boiling, in contrast to the salts with chlorides and sulphates which produce permanent hardness. Hardness is usually expressed in terms of the equivalent quantity of Calcium carbonate. Water Quality Association has provided a general scale of hardness (Table 73) demarcating the limits for soft and hard waters.

Table 73. General scale of hardness (Lehr et al, 1980).

Hardness of Water mg. L^{-1} as $CaCO_3$	Description
0-17	Soft
17-60	Slightly hard
60-120	Moderately hard
120-180	Hard
More than 180	Very hard

Much of the concern for hardness is for the problem in the domestic and industrial use rather than for its health effects. The hard water is not suitable for bathing, washing, cleaning and laundering as it precipitates the soap. In boiler feed or condenser water hardness produces heat retarding scales on the equipments, thus, resulting in higher energy cost of the industry. Hard water is not suitable for dyeing in the textile industry, beer industry and food industry.

(c) Acidity

Acidity in natural waters is caused by free Carbon dioxide ar by mineral acids. Carbon dioxide can remain in water at pH range between 4.5 and 8.5, with the values usually declining at higher pH due to conversion of some Carbon dioxide into bicarbonates and carbonates. At pH 4.5 all the Carbon dioxide remain in the unbound state indicating that this is the lowest pH that can be achieved by the presence of Carbon dioxide alone. These changes in pH due to Carbon dioxide are, however, largely governed by the biological activity such as photosynthesis and biodegradation of organic material involving bacterial respiration.

The acidity which may be considered of greater significance as regards to pollution, is mineral acidity caused by the presence of mineral acids. These acids can decrease pH to much lower levels(below 4.5). The important sources of mineral acids are mine drainage and discharge of certain industrial effluents (metallurgical industries and synthetic organic materials).

Acid mine drainage originates mainly from the ores that cotains Sulphur or iron pyrites in the form of impurities. The

Sulphur present in these impurities, is oxidized by certain bacteria of the genus *Thiobacillus* in the presence of atmospheric oxygen and water:

$$2S + 3O_2 + 2\,H_2O = 2\,H_2SO_4$$
$$FeS_2 + 3.5\,O_2 + H_2O = FeSO_4 + H_2SO_4$$

Presence of certain salts, especially of trivalent metals, like Ferric chloride and Aluminium sulphate can also produce mineral acidity after they get dissolved in water:

$$Fe\,Cl_2 + 3\,H_2O = Fe(OH)_3 + 3\,HCl$$
$$Al_2(SO_4)_3 = 2\,Al(OH)_3 + 3\,H_2SO_4$$

While Carbon dioxide acidity has got hardly any health significance, the mineral acidity can pose several environmental and health problems. Water with mineral acidity are so unpalatable that it is virtually impossible to consume them.

Mineral acidity in natural water interfere with the recreational pursuits. High concentration of Sulphuric acid causes corrosion of structures, deterioration of fisherman's nets, and eye irritation to swimmers.

Acid waters in contact with stone, concrete and metals, corrode them, solubilizing sevral heavy metals.

(d) Alkalinity

Alkalinity of water is a measure of its capacity to neutralize acids. The natural alkalinity in water is usually imparted by the salts of weak acid. Among these bicarbonate is a dominant chemical along with minor quantities of carbonate, borate, silicate, phosphate and the salts of humic and fulvic acids. Organically polluted waters may also have alkalinity derived from the salts of acetic acid, propionic acid and hydrosulphuric acid. Ammonia and hydroxides are also important sources of alkalinity in certain conditions.

A few industrial effluents are also rich in alkalinity producing substances. Cement industry wastes contains usually a very high pH (often greater than 11) due to the presence of Calcium hydroxide. Sodium hydroxide is present in waste water of several industries like soap manufacturing, textile dyeing, rubber reclaiming and tanneries.

Alkalinity is also governed in natural water by photosynthesis and microbial decomposition. Photosynthesis consumes free and half-bound Carbon dioxide leading to the formation of carbonates which in turn elevate the pH, even to a level of more than 10.

$$2\,HCO^-_3 - CO_2 + CO^{--}_3 + H_2O$$

(e) Nitrates and nitrites

Most natural waters are deficient in nitrate having a concentration usually below 5 mg/l, but certain polluted surface and ground waters may have substantially higher quantities. Its origin in water can be both from natural and man-made sources, such as fixation of atmospheric nitrogen by certain micro-organisms, plant debris, animal excreta, nitrogenous fertilizers and discharge of sewage and industrial effluents. However, its enrichment by man-made sources remains at much greater magnitude in comparison to natural ones. The agricultural sources are primarily dangerous in the farming communities using well waters. The nitrates usually remain loosely bound with soil to make them highly labile, and large quantities are thus leached out from the soil to contaminate waters, especially ground water. Very high level of nitrate have been found in ground water of several countries including India. Trivedi et al (1988) observed greater quantities of nitrates in the well waters of Maharashtra.

Nitrite is highly unstable. It is formed in natural waters mostly biochemically during denitrification and nitrification of nitrogenous compounds. Nitrite may also enter the aquatic systems through certain industrial effluents. It may also form in the distribution system by the biochemical conversion of ammonia.

Nitrate is one of the several inorganic pollutants contributed by nitrogenous fertilizers, organic manures, human and animal wastes and industrial effluents through the biochemical activities of micro-organisms. Excessive use of nitrogenous fertilizers in agriculture has been one of the primary source of high nitrate in groundwater (DeRoo, 1980; Schepers et al, 1984). Apart from nitrate, nitrogen is applied in ammonia

(NH^+_4) and amide (NH^-_2) forms, which generate nitrate nitrogen in soil systems through mineralization, which is fairly rapid in tropical and subtropical soils. Due to its high solubility in water and low retention by soil particles, nitrate is prone to leaching to the subsoil layers and ultimately to the groundwater, if not taken up by plants of denitrified to N_2O and N_2. The arrival of nitrate to ground water can be enhanced by shallow groundwater table, excessive application of nitrogenous fertilizers, manures and irrigation and abundant rainfall. Livestock feedingm, burnyards, septic tanks, animal and human contamination are the other important sources contributing high quantities of nitrate to groundwater (Stevenson, 1986). Different organisations and countries have set standards for nitrate in potable water (Table 74), to safeguard public health from the hazards associated with high concentration of nitrates.

Table 74. Maximum permissible limits of nitrate in potable water prescribed by different countries and organizations

Country/ Organization	Concentration as NO_3-N (mg/l)	Concentration as NO_3 (mg/l)	Remarks
* WHO	10	45	Latest guideline (1987)
* US Environmental Protection Agency	10	45	Max. concentration level
* ICMR (India)	10	45	IS:10500(1983)
* Canada	10	45	
* Poland	10	45	
* EEC[a]	11.30	50	Mandatory limit
* Bulgaria	6.7	30	Guide level
Belgium	11.3	50	Ref: EEC, 1977
Denmark	11.3	50	Ref: WHO, 1985
Finland	6.8	30	Ref: WHO, 1985
Hungary	9.0	40	Ref: WHO, 1985
UK	11.3[b]	50	Ref: EEC, 1977
	22.6[c]	100	
USA	10	45	Ref: EEC, 1977

[a] EEC countries are entitled to the above rules in special situation
[b] EEC Directive
[c] Chief Medical Officer's recommendation
* Source: Lunkad (1994)

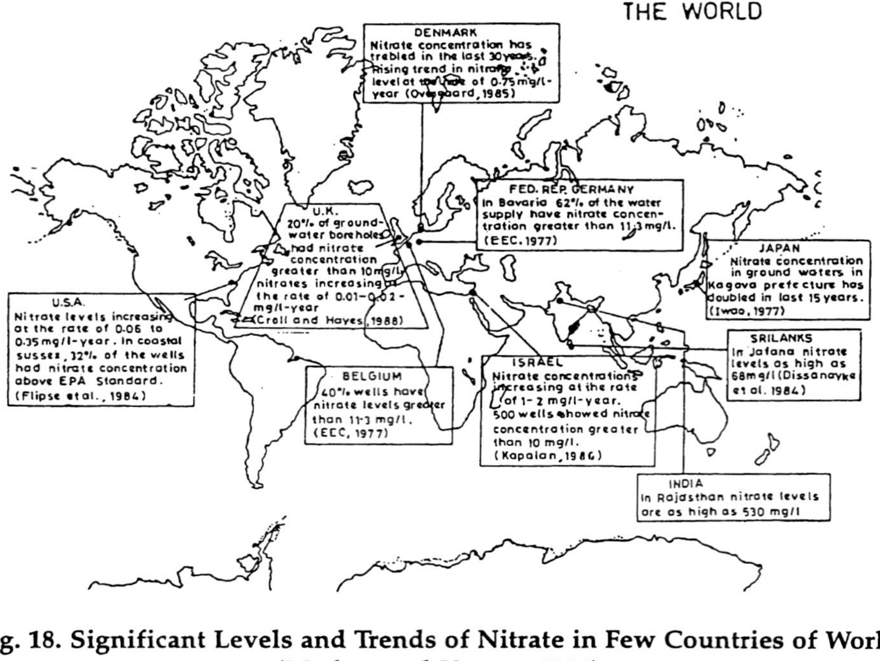

Fig. 18. Significant Levels and Trends of Nitrate in Few Countries of World (Mathur and Kumar, 1990)

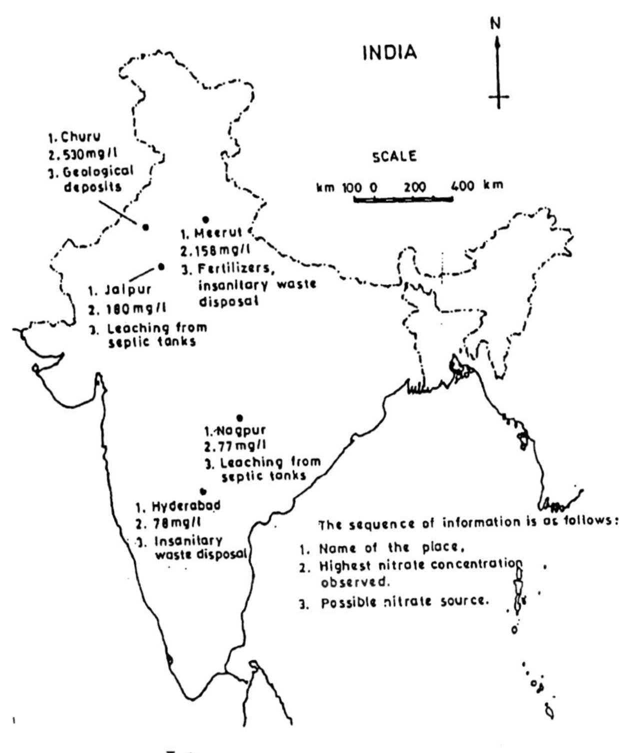

Fig. 19. Nitrate Levels in Few Cities of India
(Handa et al, 1982, Maxmanan et al, 1986; Pandey et al, 1979)

Fig. 20. Nitrate levels in Rajasthan.
(Source : Central Ground water Board, Jaipur, 1996)

Very high nitrate levels in surface waters and groundwaters have increased substantially over the last 30 - 40 years and especially in the last 20 years. Individual shallow wells in agricultural areas are particularly prone to contamination with nitrates. The rising trend in groundwater nitrate level is likely to continue for several decades, even if nitrate leaching from soils is reduced by changes in agriculture practices. Nitrate levels and trends in few countries have been shown in Figure 18. Figure 19 and 20 show nitrate levels in India and Rajasthan respectively.

Nitrate levels in ground water for different concentrations ranges in few districts of Rajasthan (Table 75) show that 74 percent groundwater samples in Barmer and 40 percent in Jaipur district exceeded nitrate concentration of 0.3 mg/l.

Table 75. Percentage frequency of groundwater nitrate for different concentrations ranges in few districts of Rajasthan (Gopal et al, 1983; Gupta, 1981).

District	Nitrate concentration range (mg/1)			
	0-4.5	4.5-11.3	11.3-22.6	22.6
Barmer	18	8	11	63
Bikaner	66	27	5	2
Jaipur	38	22	17	23
Jaisalmer	50	19	9	22
Udaipur	47	31	14	8

Nitrate content in groundwater of Varanasi (Table 76) show nitrate content in dugwell and handpump. The nitrate content was highest in handpump at Kamauli (342 mg/l). In dungwell it was maximum (284 mg/l) at Shivpur and minimum (59 mg/l) at Ashapur and Chiraigaon. High nitrate in ground-water samples were assumed to have developed due to seepage of urban sewage from a 400 km long old sewarage system which had developed leakage at several places. High nitrate was also found around an unlined wastewater carrying surface drain near Banaras Hindu University.

Table 76. Nitrate content in groundwater of Varanasi

Sector of city area	Location	Groundwater structure		NO_3 level
		Type	Depth (m)	(mg/1)
Northern	Shivpur	DW	10.35	284
	Paignambarpur	DW	17.00	62
	Ashapur	DW	19.20	59
	Chiraigaon	DW	15.30	59
	Kamauli	HP	40.00	342
		DW	20.00	99
Central	Alaipura	DW	13.90	66
	Bhojpur	DW	16.80	199
	Shivdaspur	DW	16.70	128
	Lohta	DW	15.60	180
Southern	Kakarmatha	DW	20.50	100

DW : Dug Well HP : Hand Pump

Source : Ground Water Pollution Directorate, 1993

Nitrate in groundwater in Delhi (Table 77) indicate that the shallow water bearing high nitrate concentration were unfit for drinking. The nitrate levels was maximum in Delhi zoo area and minimum in New Friends Colony. The groundwater in and around Delhi exceeded maximum permissible limit for drinking water (100 mg/l) set by BIS.

Nitrate concentration in drinking water is considered important for its adverse effect on human health. Nitrate itself is relatively non-toxic, but when ingested in food or water, it may be reduced to nitrite by bacteria in the mouth and gut. Nitrite is a powerful oxidising agent and converts Iron in the haemoglobin from ferrous to ferric state and due to this haemoglobin loses its property to carry oxygen. Nitrite may also react in-vivo with nitrostable substrate in ceratin foods to form N-nitroso compounds which are suspected to be carconogenic. The association between infant methaemoglobinaemia and the consumption of water with high nitrates has been established (Shuval and Gruner, 1977). Most cases of methaemoglobinaemia have been associated with nitrate levels exceeding 25 mg/l, but in some cases nitrate level was only 10-20 mg/l, beside individual susceptibility.

Table 77. Groundwater nitrate content in some areas of National Capital Territory of Delhi

Areas	NO_3 (mg/l)	Areas	NO_3 (mg/1)
Block - Kanjhawala		Bokali	113
Kanjhawala	140	Mamurpur	236
Mohammad Pur Mairi	161	*Block - City*	
Mundka	119	India Gate Nursery	121
Kirari	184	Delhi Zoo	1589
Sultanpuri	208	Jamia Milia Market	194
Mangol Pur Khurd	114	Vasant Enclave	133
Mangol Pur Kalan	652	Hastal	171
Nijampur	106	Parsi Dharam Shala	100
Mangolpuri	180	Naharpur	108
Bhagey Bihar	115	Jeevan Nagar	232
Jat Khor	390	Kishan Ganj	190
Krishna Bihar	184	Kotla Mubrakpur	328
Block - Najafgarh		Sanjay Camp	240
Palam Gaon	105	Sudershan Puri	202
Chawala	130	Shanti Van	144
Kangan Hari	300	Okhla Ind. Phase I	110
Kharkhari Nahar	203	Sarita Vihar	149
Kakrola	379	New Friends Colony	102
Pindawala Kalan	361	Naraina	600
Dhansa	377	Begampur	240
Ujwa	508	*Block - Shahdara*	
Jharoda Kalan	208	Ghazipur	139
Dichaun	178	East Vinode Vihar	180
Neejwal	117	Durga Puri Chowk	141
Gazipur	389	*Block - Mehrauli*	
Kair	264	Dera Gaon	690
Isapur	106	Rajokari Road	262
Block - Alipur		Gadaipur	743
Puthkhurd Village	266	Jaunapur	124
Palla	277		

Source : Central Ground Water Authority (Times of India, 16.11.98)

(f) Flourides

Earth's crust contains abundant fluoride as in high Calcium granite (520 mg/l), low Calcium granite (850 mg/l), alkaline rocks (1200-8500 mg/l), Shales (740 mg/l), sandstone (270 mg/l), deep sea clays (1300 mg/l), and in deep sea carbonates (540 mh/l) (Bulusu et al, 1985). India is considered to be one of the richest countries in the world for the occurrence of fluoride bearing minerals. The natural occurrence of fluoride is generally limited as most waters have a concentration lower than 0.5 mg/l. However, industrial activities can increase its concentration significantly in the water. Fluoride compounds are widely used in several industries such as Aluminium works, bricks, ceramics, insecticides, chemicals and fertilisers that lead to the contamination of water in the vicinity of these areas. The combustion of coal also introduces fluorides in the environment.

Fluoride with lower concentration at an average 1 mg/l is regarded as an essential constituent of drinking water mainly because of its role in prevention of dental caries (McClure, 1970). High incidence of dental caries have been reported where fluoride concentration usually remain lower than 0.5 mg/l in drinking water.

However, in many areas the natural water are quite rich in fluoride and their consumption by cattle and human beings has resulted in Fluorosis. Examples of such areas in India can be cited from certain parts of Andhra Pradesh and Rajasthan where the Fluorosis occurs quite commonly due to fluoride which remains much higher than the prescribed limits in ground waters.

Singh et al (1962) and Kanwar and Mehta (1968) reported cases of endemic fluorosis in human beings caused by the ingestion of high fluoride well water in Haryana and Punjab. Siddiqui (1972) reported fluoride content of water in Andhra Pradesh. Some 600 persons among 21,000 in 28 villages of the three districts of Andhra Pradesh - Prakasam, Nalgonda and Guntur have been crippled for life by fluorosis. It has been estimated that 20-25 million Indians are afflicted by fluorosis.

Rajasthan's fluorosis problem has reached threatening proportions, fluoride has permanently crippled over 3.5 lakh inhabitants and likely to cripple many more. In some cases, due to the compression of nerves by awkwardly growing bones, paralysis sets in. Fluordisis is prevalent in the districts of Jodhpur, Bhilwara, Jaipur, Bikaner, Udaipur, Nagaur, Barmer and Ajmer. Banka Patti (Nagaur) is among the several villages in the state where fluoride pollution poses serious health hazards. Rajasthan is believed to have the maximum number of the humped back because of high fluoride concentration in water sources in arid and semiarid zones.

In Rajasthan more than 80 percent of the districts (Figure 21) have high fluoride levels (12-90 ppm in Thar region and 1.44-28.1 ppm in Aravalli hill region) in their drinking/ ground waters (Teotia and Teotia, 1984; Susheela, 1993) and distributed heterogeneously. The first case of fluoride toxicosis (skeletal fluorosis) was reported from Jaipur district (Kasliwal and Solomon, 1959). A few reports on endemic fluorosis have also been reported from Jhunjhunu (Thergaonkar and Bhargava, 1974) and Ajmer (Mathur et al, 1976). In some areas of Dungarpur district maximum deposition of fluoride rocks and some fluoride mines are still running. These areas have relatively a high fluoride concentration (1.2 to 8.9 ppm) in their drinking waters.

In Rajasthan fluoride is heterogeneously distributed in 28 districts (87.5%) out of 32. From its Thar region 12-90 ppm fluoride in drinking water/ground water has been reported and the highest fluoride level (90.0 ppm) was found in Nagaur district followed by Churu (32 ppm), Ganganagar (26 ppm), Bikaner (22 ppm), Barmer (19.6 ppm), Pali (18.3 ppm) and Jhunjhunu (12 ppm). Although fluoride content is also found in Aravalli hill region, but lower than that in the Thar region. The maximum (28.1 ppm) fluoride content so far reported from this region was in Jaipur district followed by Bhilwara (24 ppm), Udaipur (21.6 ppm), Bharatpur (18.4 ppm), Ajmer (16.2 ppm), Dungarpur (10.8 ppm), Bundi (6.8 ppm), Chittorgarh (6.6 ppm), Banswara (5.5 ppm), Kota (4,8 ppm), Rajsanand (4,5 ppm), Alwar (4.0 ppm), Tonk (4 ppm), Baran (2 ppm) and Jhalawar (1.4 ppm).

Purohit (1988) observed that well waters of several vill villages of Nagaur and Churu districts contain 60-100 ppm fluoride. Sampling of underground water from Rajasthan reveals

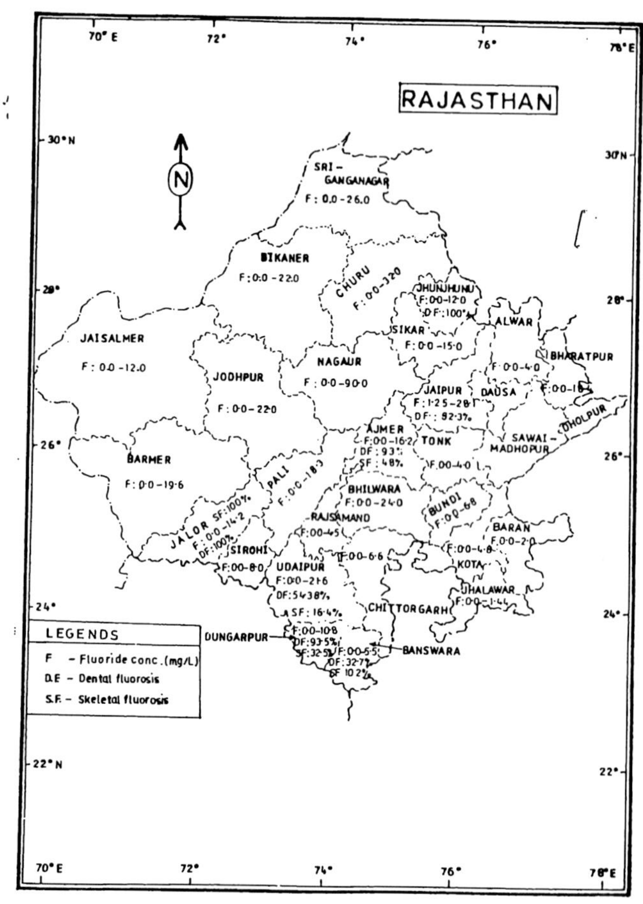

Fig. 25. Fluride and Fluorosis in Rajasthan

that out of 26,778 villages surveyed, 5659 contained 1.1-2.2 mg/l, 2158 contained 2.1-3.0 mg/l, 1308 contained 3.0-5.0 mg/l, 515 contained 5.1-48.0 mg/l.

In Jaipur district's Phagi tehsil's 18 villages were surveyed for fluoride content in well waters, 8 villages showed excess fluoride beyond permissible limit (1.5 mg/1), Renwal Mangi 3.4 ppm, Thal 3.7 ppm, Chitora 1.9 ppm, Panjadi 1.9 ppm, Madhorajpura 3.6 ppm, Jankinathpura Dola Aali 9.4 ppm, Gulabpura 1.9 ppm and Heerapura 6.2 ppm. In Chomu tehsil out of 18 villages surveyed only 2 villages Malikpura and Neendhra had fluoride in drinking water 2.7 ppm and 2.0 ppm respectively. From Phulera out of 6 villages surveyed 2 village Mohan-ka-bas and Pachkodia had fluoride content 4.1 and 6.2 ppm respectively (Saxena and Nathawat, 1989).

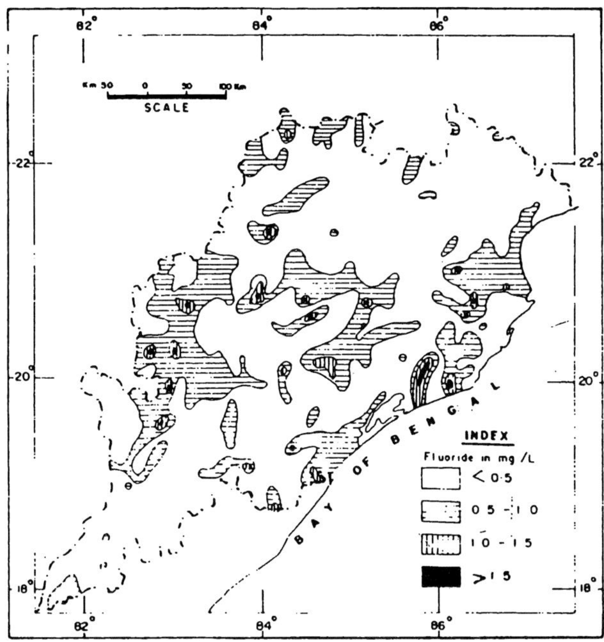

Fig. 22. Distribution of Fluoride in Orissa

Table 78. High values of Fluoride in Dugwell waters (shallow aquifers) in Orissa

District	Location	Fmg/L	Location	F mg/L	Location	F mg/L
1	2	3	4	5	6	7
Balasore	Birnarsinghpur	2.4	Chatrang	3.4		
Puri	Gargh Rupass	4.9	Jageswarpur	3.0	Astarang	2.7
	Konark	3.0	Jiunti	3.7	Marad	5.9
	Kujamathar	1.9	Kuans	13.0	Jariparaha	2.9
	Puri	2.5	Aunlo	7.4	Singhpur	8.2
	Balakati	2.8	Kualo	2.8	Brahamapur	16.4
	Nuagaon	4.5	Sarakana	1.5	Alanda	4.7
	Kasarda	7.5	Kulantira	4.0	Algum	1.5
	Puri town	4.8	Sasan		Charsi-I	3.0
	Budiabar	4.0	Badagualipara	1.9	Charsi-II	2.9
	Rabanugaonn	7.0	Narakhandi	2.3	Alipara	2.6
	Gopinathpur	1.7	Kanas	2.1	Bhudbar	3.6
	Mahipur	4.3	Kasarda	5.3	Baharma	10.5
Sambalpur	Balitikra	1.6				
Sundergarh	Subdega	1.6	Motto	8.3		

1	2	3	4	5	6	7
Bolangir	Chatta makhan	2.6	Khaprakhol	2.2	Khari	1.9
	Deogaon	3.5	Jogimunda	1.9	Sonepur	1.7
Cuttack	Karandibandh	1.9	Barapank	1.8	Palai	3.2
	Rajnagar	1.6	Kasoti	1.5	Barayula	3.2
	Nahanga	1.7	Raghunathpur	2.3	Adhang	1.5
	Matia	1.8	Balituth	6.9	Muradpur	2.7
	Nembal	2.5	Bari	4.6	Badaraja	3.4
	Bhutmundai	1.9	Sonagopalpur	1.7	Katana	2.9
Dhenkanal	Thakugarh	2.9	Bantala	1.5		
Ganjam	Rambha	1.5	Tumbagarh	1.5		
Kalahandi	Golamunda	2.5	Utkela	5.0	Boden	2.2
	Santapur	1.9	Koksara	2.6	Dharambandh	3.2
	Tundla	2.3	Ranmal	3.1		
Keonjhar	Ranchipur	2.6				
Phulhabi	Charichak	3.0	Nuagaon	3.8	Dandua	2.2

Table 79. Some high values of fluoride in Tubewell waters (deeper aquifers) in Orissa

District (undivided)	Location	F mg/L	Location	F mg/L	Location	F mg/L
Balasore	Defence colony	3.2	Jco mess	3.1	DRDO Guest	2.0
	Officers colony	2.5				
Bolangir	Badmal	2.2	Lohapalli	1.5	Lathora	1.8
	Dhanmunda	2.4	Khaprakhol	3.6	Binka	1.9
	Kanut	1.7				
Kalahandi	Golamunda	1.5	Lakhana	2.7	Ladugaon	1.7
	Ranmal	1.8				
Keonjhar	Panndapada	2.4	Jhanghira	2.4		
Puri	Jariparha	2.6	Satgaragaon	2.1	Sagaragaon	3.4
	Singhpur	2.0	Bajrakot	8		
Phulbani	Tikabali	2.5	Chohipata	1.7	Damamunda	2.2
	Karnkata	3.6	Kantamal	1.6		
Sundergarh	Sargipalli	2.0	Ekma	1.5	Talsara	3.2

The distribution of fluorides in the ground water in shallow aquifers in Orissa belong to both hard rocks and the coastal alluvium (Figure 22)

Fluoride content of the ground water was low (Table 78). Ganjan, Karaput and Mayurbhanj districts showed fluoride in the concentration of 1.5 mg/l. Bolangir, Kalahandi, Phulbani, Puri, Sambalpur and Sundargarh had fluoride content of more than 1.5 mg/l, which might have been derived from weathering of the crystallines.

Generally, tubewell (deeper) waters in Orissa were found to contain less than 1.0 mg/1 fluoride. The higher values found in some cases have been shown in Table 79. Hot spring waters in Brahmagiri and Phulbani districts and Tarbol in Puri district were found to contain 9.2 and 12.2 mg/l of fluoride respectively.

The permissible level of fluoride according to WHO (1984) in drinking water is 1.5 ppm (1 mg/l). The optimal concentration of fluoride in drinking water at which baneficial effect of fluoride is observed is 0.7 and 1.2 ppm. Higher fluoride concentration do not yield better results.

Fluoride first attack the teeth, deforming them beyond recognition. Prolonged intake of fluoride containing water stiffs the bones joints, particularly those of the spinal cord. The hump once developed can not be cured.

Fluoride in not absorbed in the blood stream. It has an affinity for Calcium and gets accumulated in the bones, resulting in the mottling of teeth, pain in the bones and outward bending of legs from the knees (Knock-knee syndrome).

Fluoride concentration in about 41,083 villages of Rajasthan was reported to be greater than the maximum permissible level of 1 mg/l of water, it ranged between 1.1 to 2 ppm in about 2158 villages, and 10 ppm to 30 ppm in atleast 48 villages.

Cattle grazing around fluoride pollution sources, such as Ceramic works, Phosphate fertilizer works, and Aluminium factories often develop fluorosis (Schmidt and Rand, 1952). The toxicity symptoms developed involve staining, mottling and abrasion of teeth, increased concentration of fluoride in urine and bone, temporary or permanent lameness and decreased milk production in dairy cattle. The severity of these symptoms will depend on the fluoride content of vegetation and the duration of ingestion. Consequent of these disorders, the animals become lethargic and emaciated and their productive life is considerably reduced (Little, 1970). Presence of fluoride in the developing teeth interferes in the dental structures resulting in the typical brown to blackish stains on the enamel and the affected tooth erupts with characteristic markings of mottling, staining, hypoplasis and/or hypo-calcification (Shupe and Alther, 1966). These interferences may cause incomplete development of enamel, dentine or the entire tooth. Excess fluoride results in ligament calcification and the production of bony outgrowths that frequently lame the animal.

Fluoride is the most exclusive bone seeking element owing to its affinity for Calcium phosphate. It is the componeat part of bones and teeth tissue contains about 0.02 percent of fluoride. Its major part being in tooth enamel, whose composition is close to the formulae $CaF(PO)_4$. Soluble fluoride in the human body is absorbed almost completely (86-97 %) regardless of the concentration in the drinking water (Irving, 1974). The Indian Council of Medical Research (1975) has given the highest desirable limit of fluoride as 1.0 mg/1 and maximum permissible limit as 1.5 mg/1. Excess fluoride exerts various adverse effects on human body (Table 80).

Table 80 . Effect of excess fluoride on human body.

F mg/L	Physiological Effect
1.0	Dental carries reduction
2.0	Mottled enamel
5.0	Osteoscleorosis
8.0	10% Osteoscleorosis
20-80	Crippling Fluorosis
100	J Retardation
125	Kidney changes
2500	Death

The optimal fluoridation of water samples at about 1 ppm of fluoride decreased the "delayed missing field" rate by about 60 percent in children who consumed the treated water during the period of tooth formation. High level of fluoride found in the dental plaque inhibit the enzymes of the bacteria in, the plaque. These enzymes are necessary for the conversion of sugars into the acid which cause tooth decay. Low levels of fluoride ingestion during the period of tooth formation can also cause "mottling" of teeth. In its mildest form, mottling is a fairly innocuous cosmetic effect in which opaque, paper white areas are seen on the surface of normally translucent teeth. 1-2 percent of the people using water containing 1 ppm fluoride had atleast 2 teeth with 25-50 percent of their surface affected by such mottling. As the level of fluoride goes up, brown stains appear on the teeth. Such brown stains begin to appear in the population when water contains 1.8-1.9 ppm fluoride. Higher fluoride levels can cause severe discolouration pitting and eventually erosion of the tooth enamel. Fluoride readily affects the teeth only during enamel formation upto about the age of 12 years in human beings.

The prevalence of severity of dental fluorosis was found to be increased with increasing fluoride levels in water and age (Table 81 and 82). The maximum prevalence (99.2 percent) of dental fluorosis was observed in 17-22 year age group, and the minimum (83.3 %) in the 5-10 year age group. The appearance of dental fluorosis in adults was slightly different from that of children. The anterior teeth of adults showed diffused form of dental fluorosis. None of the children showed evidence of a further stage of dental fluorosis.

Table 81. Prevalence (%) of dental fluorosis in relation to fluoride levels in Dungarpur district of Rajasthan (Choubisa et al, 2001)

	Village	Mean fluoride conc (ppm)	Questi-onable	Very mild	Mild	Mode-rate	Severe
1	2	3	4	5	6	7	8
1.	Gokulpura	1.7	19.1	25.9	12.9	8.6	1.8
2.	Dovada	2.8	14.7	16.3	21.3	14.6	8.9

1 2	3	4	5	6	7	8
3. Hathai	3.3	6.9	21.4	26.4	20.9	10.4
4. Kahela	3.4	6.4	19.9	27.9	27.4	11.8
5. Pratappura	3.6	5.1	15.9	29.9	31.9	14.4
6. Mandav	3.8	4.7	13.1	26.7	35.1	18.8
7. Nalwara	3.8	4.2	11.6	24.7	37.4	21.0
8. Bhavanpura	4.8	3.8	10.2	25.0	37.2	23.1
9. Gehuwara	5.8	2.5	9.0	23.5	39.5	24.5
10. Samota	6.1	1.9	8.3	21.6	41.2	26.9
Total		6.9	15.0	24.2	29.7	16.4

Table 82. Prevalence (%) and severity of dental fluorosis in relation to age in Dungarpur district of Rajasthan (Chaubisa et al, 2001).

Age group (Years)	Questi-onable	Very mild	Mild	Moderate	Severe
5-10	9.25	25.9	23.1	18.9	6.01
11-16	11.85	21.9	25.0	26.6	5.59
17-22	5.8	19.5	26.9	38.5	8.3
23-28	9.9	20.4	24.8	28.0	11.8
29-34	10.3	20.6	24.4	28.2	10.8
>34	4.9	12.9	20.2	36.2	16.8
Total	8.4	19.2	23.6	30.3	10.5

Observations on prevalence of dental fluorosis among school children in five schools of village Juaikalan, Bhiwani, district, Harayna (Table 83 and 84) show that 100 percent prevalence of fluorosis was observed among the male children of above 15 years age group. In age group of 11-15 years, 194 cases of dental lession were observed out of total of 198 individuals. In this category 52.06 percent were found at yellow brown stage, which was highest among all the three stages. A total of 258 out of 272 males were observed to be effected by dental fluorosis in the age group of 6-10 years. Dominant stage of this problem was recorded as Chalky white (44.57 %) and spots type (47.67 %). Minimum cases of fluorosis were observed in male children of age group 1-5 years. Yellow brown stage and spot were dominant in this age group.

In the females 97.50 percent cases of dental lesions were observed in the higher age group (11-15 years), on the same pattern as in case of male . In this age group 44.87 percent cases were observed in yellowish brown stage and 41.02 percent in horizontal streak type. In the age group of 6-10 and 1-5 years, the percentage of dental fluorosis was 89.56 and 76.27 percent respectively. In the age group of 6-10 years 65.03 percent cases were in Chalky white stage whereas a maximum of 45.55 percent was observed in yellowish brown stage among 1-5 years age group.

Further it was observed that the prevalence of dental fluorosis in male children was higher than in female ones.

People residing in high fluoride places over a period of time manifest "skeletal fluorosis". The disease is known to affect vertebral column, pelvic girdle and ribs. Maximum changes are detected in vertibral column particularly in the cervical and lumber region. Fusion of vertibral bodies may occur, resulting in neurological symptoms and severe neck stiffness, these being the commonest findings among the patients of fluorosis.

In high fluoride areas, the intake of contaminated water cause bone growth that cause severe skeletal fluorosia. Most of the clinically significant bone abnormalities caused by fluoride appear in the spine. Pointed protrusions grow out of the vertibras and tend to lock one vertibrae to another. The inability to bend the "pocker back" are characteristic of this severe skeletal fluorosis. Extra bone growth on and between the ribs make breathing difficult. Excessive fluoride intake can bring skeletal fluorosis past even the rigid and pocker back stage.

Evidence of skeletal fluorosis (Table 85) of varying grade was noted in boys (3.9 %) , girld (1.7 %), males (28.4 %) and females (23.9 %). The highest skeletal fluorosis prevalence (37.7 %) was found in adult male at 6.1 ppm and the lowest (7.7 %) in female children at 6.1 ppm of fluoride. The prevalence of skeletal fluorosis also increased with increase in fluoride concentration and with age. Males showed comparatively higher prevalence of

Table 83. Incidence and Severity of Dental Fluorosis Among male School Children at Village Juai Kalan, Distt., Bhiwani. (Dahiya et al, 2000).

S.No	Age Group in yrs	No. of Indivi-duals Examined	No. of DF* (+ve) affected individuals	Stage of DF*			Type of DF*		
				Chalky white	Yellowish brown	Brownish black	Horizontal streaks	Spots	Both
1.	1-5	161	130 (80.75)	56 (43.07)	59 (45.38)	15 (11.54)	34 (26.15)	66 (50.77)	30 (23.08)
2.	6-10	272	258 (94.85)	115 (44.57)	110 (42.64)	33 (12.79)	69 (26.74)	1223 (47.67)	66 (25.58)
3.	11-15	198	194 (97.98)	73 (37.63)	101 (52.06)	20 (10.31)	75 (38.66)	48 (24.74)	71 (36.60)
4.	>15	3	3 (100)	2 (66.66)	0 (0.00)	1 (33.33)	1 (33.33)	0 (0.00)	2 (66.66)

* DF = Dental Fluorosis

Figure in parenthesis indicate percentage.

Table 84. Incidence and Severity of Dental Fluorosis Among Female School Children at Village Juai Kalan, Distt., Bhiwani. (Dahiya et al, 2000).

S.No	Age Group in yrs	No. of Indivi-duals Examined	No. of DF* (+ve) affected individuals	Stage of DF*			Type of DF*		
				Chalky white	Yellowish brown	Brownish black	Horizontal streaks	Spots	Both
1.	1-5	118	90 (76.27)	39 (43.33)	41 (45.55)	10 (11.11)	31 (34.44)	44 (48.89)	15 (16.67)
2.	6-10	182	163 (89.56)	106 (65.03)	47 (28.83)	10 (6.13)	49 (30.06)	66 (40.49)	48 (29.45)
3.	11-15	80	75 (97.5)	32 (41.02)	35 (44.87)	11 (14.10)	32 (41.02)	30 (38.46)	16 (20.51)
4.	>15	–	–	–	–	–	–	–	–

* DF = Dental Fluorosis

Figure in parenthesis indicate percentage.

skeletal fluorosis. The onset of skeletal fluorosis appeared in children (>16 years) at 4.8 ppm mean fluoride concentration where as, in adults, although it appeared at 1.7 ppm, but in the higher age group (>45 years). Crippling fluorosis was found at and above 3.3 ppm fluoride. The more severe forms of skeletal fluorosis-genu-varum-syndrome, invalidation and kyphotic deformities were observed only in adults over the age of 48 years who had drunk water containing 3.3 ppm fluoride. Secondary neurological complications such as quadriplegia were only seen in adults (>55 years) drinking water with 3.8 ppm fluoride.

Table 85 . Prevalence of skeletal fluorosis (%) in children and adults of Dungarpur district of Rajasthan (Chaubisa et al, 2001)

Village	Mean fluoride	Children			Adults		
	cnc (ppm)	Male	Female	Total	Male	Female	Total
1. Gokulpura	1.7	0.0	0.0	0.0	11.4	6.9	9.4
2. Dovada	2.8	0.0	0.0	0.0	13.7	11.1	12.5
3. Hathai	3.3	0.0	0.0	0.0	18.2	14.7	16.4
4. Kahela	3.4	0.0	0.0	0.0	21.0	18.0	19.6
5. Pratappura	3.6	0.0	0.0	0.0	31.6	21.4	26.7
6. Mandav	3.8	0.0	0.0	0.0	28.1	21.4	25.5
7. Nalwara	3.8	0.0	0.0	0.0	31.1	27.8	29.4
8. Bhauanpura	4.8	10.0	0.0	5.2	36.4	34.6	35.6
9. Gehuwara	5.8	9.5	8.0	8.7	43.4	35.9	39.6
10.Samota	6.1	18.5	7.7	13.2	48.7	43.8	46.3
Total		3.9	1.7	2.9	28.4	23.9	26.2

(g) Cyanides

Cyanide ion (CN^-) concentration in waters results primarily from the industrial wastes and use of pesticides. The effluent containing significant quantities of cyanide arise from electroplating industry, metal finishing works, photo processing, refining and coal coking units.

The presence of other pollutants with cyanide in natural conditions can modify its toxicity to a great extent. For example

Copper is antagonistic (Mason, 1981), while Zinc and Cadmium are synergistic to the cyanide toxicity. The toxicity of cyanide is also increased in the conditions with high temperature and low oxygen in water.

The limit for cyanide in drinking water has been kept from 0.01 mg/1 to 0.05 mg/1; Fortunately it does not accumulate for logger periods in water owing to its degradation by certain bacteria who can use it for their enrgy requirements.

(h) Toxic gases

Gases in water may be important when they produce taste and odour or are corrosive or pose explosion hazard. Hydrogen sulphide and methane are two common gases of concern which frequently exist in water.

Hydrogen sulphide commonly originates in waters owing to the decomposition of organic matter. Many ground waters have significant quantities of hydrogen sulphide arising from the decomposition of naturally occurring organic matter. Sewage and certain industrial effluents have considerable quantities of sulphur consuming organic materials whose discharge in waters promote anaerobic conditions and accumulation of Hydrogen sulphide. Excessive quantities of Hydrogen sulphide produced in waters may release some gas into the atmosphere giving a typical rotten egg odour.

Methane is produced in natural waters mostly from anaerobic decomposition of organic matter. It also orginates in the effluents of certain industries like Petroleum refining and natural gas. The presence of Methane in certain ground waters is ascribed mainly to the decomposition of naturally occurring organic matter in anaerobic conditions.

Major hazards associated with the presence of Methane are related to its explosive nature and the danger of asphyxiation. Water supplies that contain Methane should thus, be carefully vented so as to avoid its accumulation in pipes, water heaters, pressure tanks and water treatment equipments.

Chapter-5

ORGANIC POLLUTION

Organic pollution of water refers to the pollution caused by organic materials that can be readily decomposed by the micro-organisms. All natural bodies of water oxidise organic matter without development of nuisance conditions provided that the organic matter loading is within the range of oxygen resources of the water. In balanced aquatic ecosystems the indigenously produced organic matter is always in quantities that allows its complete degradation by the oxygen available in water, without disturbing its natural balance. It is only when the system receives organic matter in excess from outside, the oxygen tend to diminish and lead to the development of hypoxic or totally anaerobic conditions.

Both anaerobic and aerobic organisms are ubiquitous in nature. An increase of the organic matter exhausts the oxygen supply of waters, and the activity of aerobic organisms. While the aerobic conditions usually result in the formation of innocuous oxidized end products, the anaerobic conditions promote the reduction of various inorganic chemicals, notably nitrates and sulphates, along with the accumulation of several obnoxious compounds.

The overall decomposition of organic matter yields stabilized end products including the plant nutrients leading to the eutrophication, especially in the regions away from the mixing of the organic wastes. All these changes caused by organic pollution bring about adverse effects on the aquatic organisms, and degrade the quality of water to render it unfit for a variety of uses.

NATURE OF ORGANIC MATTER

All the organic substances which can serve as food for living organisms to obtain their energy needs and to build new cells can be termed as organic matter. The organic matter based on its origin, can be divided into two main categories, that is, synthesized naturally by biochemical means, and synthesized chemically in the laboratory. However, except for a few, most synthetic compounds are biologically inert to degradation and, hence, cannot be considered in the category of organic matter.

The most commonly encountered organic material, both in the nature and in organic wastes, are the biologically originated carbohydrates, fats (lipids) and proteins, which are readily utilized by the micro-organisms as their energy sources, and even serve as food for higher life including human beings. The micro-organisms, besides these three principal foods, are also capable to utilize a number of other organic substances ranging from simple hydrobarbons to complex biomolecules, but these are much less abundant than the former. Because of the dominance of these three organic materials in most waters and waste waters, they are of major concern in water quality control.

Carbohydrates

Carbohydrates are the organic compounds containing carbon, hydrogen and oxygen (CHO) in which the hydrogen and oxygen atoms are present in some ratio in which they occur in water (2 hydrogen : 1 oxygen). A number of industries which process wood, paper, cotton, textile and food have substantial quantities of carbohydrates in their waste waters. The carbohydrates can be grouped in thee main categories:

(i) Monosaccharides
(ii) diasaccharides
(iii) Polysaccharides

Monosaccharides are the simple sugars made-up of either five carbon chains (pentoses) or six carbon chains (hexoses). Some of the common examples of the former are xylose, arabinose; and the latter are glucose, fructose, galactose and mannose. The gerneral formula for pentoses is $C_5H_{10}O_5$ and for hexoses $C_6H_{12}O_6$.

Diasaccharides are composed of two molecules of hexose sugars with a general formula of $C_{12}H_{22}O_{11}$. There main types of important diasaccharide sugars are sucrose, maltose and lactose.

Polysaccharides are condensation products of Monosaccharide sugars. The environmentally important Polysaccharides are starch, cellulose and hemicellulose. Starch has a general formula of $(C_6H_{10}O_5)_X$,in which hundreds of glucose molecules are joined in a definite fashion. Cellulose also has the same general formula as starch, but the glucose units are joined differently. Hemicellulose have alternate units of pentoses and hexoses .

FATS (LIPIDS)

Fats also contain C, H and O but unlike carbohydrates, the oxygen is present in much lesser quantity. They are sparingly soluble in water but get readily dissolved in organic solvents. Chemically, the lipids are esters of glycerol. They are easily hydrolysed to corresponding fatty acids and glycerol and serve as an excellent food for a number of micro-organisms.

The fatty acids present in the lipids are usually 16-18 carbon atoms, though acids with smaller carbon chains may also occur in appreciable quantities in certain fats. The lipids are called oils when they are liquid at ordinary temperatures, and fats when they are solids. The oils have a predominance of unsaturated fatty acids, whereas fats usually have saturated acids.

Proteins

Proteins have mainly C,H,O and N with small quantities of S occasionally. Because of the high nitrogen content of proteins which varies from 15-18 percent, they are commonly considered as nitrogenous organic matter.

Chemically proteins are made up of amino acids which are linked by polypeptide chains. As many as 20 different amino acids are found in nature which form the building blocks of proteins in different sequences.

An amino acid is amphoteric in character having two active groups, the amine (- NH_2) and acid (-COOH). A few amino acids like Cysteine and Methionine also contain Sulphur. While many amino acids are of straight or branched chains (Glycine, Leucine), some may also contain aromatic (phenyl alanine, tryosine) and heterocyclic (proline, tryptophan) rings in their structure. Proteins upon hydrolysis, give varying amounts of amino acids which can be later degraded by different bacteria.

PHOSPHORUS CONTAINING COMPOUNDS

A large number of organic biomolecules of complex structure may also be present in organic matter that contain phosphorus in varying quantities, such as nucleic acids (DNA and RNA), polyribose nucleic acids, adenosine triphosphate, adenosine diphosphate, adenosine phosphosulphate, nicotinamide adenine dinucleotide, NAD-phosphate, 3-glycerophosphate, 1,3-diphospho-glycerate.

The phosphorus in these compounds is mainly present as a phosphate radical which is bound to the organic molecule by an ester bond, and can be easily separated by hydrolysis.

OTHER ORGANIC COMPOUNDS

Urea is an important compound present in appreciable quantity in the excreta of animals and human beings. It is a nitrogen bearing compound. A large number of other minor compounds, many of which are biochemical intermediates or end products of incomplete degradation (such as alcohols, acids, ketones, aldehydes, methane etc.) are also released by the organisms in the environment. These can be further degraded by the micro-organisms under suitable conditions, or can remain as refractory substances. A large number of other obnoxious organic compounds have also been found to be excreted or secreted by organisms in the water. Most of these compounds however, occur in trace quantities.

SOURCES OF ORGANIC MATTER

The most important source of organic matter in waters are the disposal of municipal sewage, industrial effluents, urban and rural run-off and the detritus formed by indigenous primary and secondary production.

The municipal sewage consists mainly of human excreta and other household organic wastes. Average raw sewage contains usually 1.0 percent solids which remain both in suspended and dissolved form. Out of these 70 percent are of organic nature and remainder are of inorganic form. Of the organic constituents, 65 percent are nitrogenous (mainly proteins), 25 percent are carbohydrates, and 10 percent are fats (Tebbut, 1977). Discharge of sewage can put great quantities of nitrogenous organic matter in the bodies of water. Even after the conventional secondary treatment, a large quantity of dissolved organic matter remains in sewage which may be still enough to cause the organic pollution, especially in the waters with low self-purification capacity.

India has the sewer facility for about 2 percent of its population (Rao, 1981), and the sewage generated in Class I cities gets some kind of treatment for only 25 percent of the total volume. Most of the small towns and cities do not have any treatment facilities and the sewage is diverted directly into the nearby water bodies.

Industrial effluents are important source of organic pollution of different nature. A number of industries like food processing, dairy, paper and pulp, sugar, distilleries, breweries, tanneries, textile and others produce considerable quantities of organic effluents of varying nature besides other pollutants (Table 86). The volume of effluents from these industries is relatively large and cause organic pollution after they are diverted to municipal sewers or are directly discharged into the aquatic systems with or without any treatment.

The run-off originating from urban areas picks-up large quantities of organic matter from the streets, houses, gardens and other places and reach bodies of water through municipal sewers or directly. The urban run-off is especially important at the onset of rainy season after the dry spell during which huge quantities of organic matter accumulate in city drains. Though the BOD of urban run-off is highly variable, its values as high as 7700 mg/l have been observed in certain cases (Mason, 1981).

Table 86. Characteristics of some important organic wastes released from different industries (Data based on Nemerow 1978)

Industry	Characteristics
Daily products	High dissolved organic matter mainly proteins, fats and lactose
Canned food	Organic matter in colloidal and dissolved form, High TSS
Meat and Poultry Product	High dissolved and suspended organic matter especially products proteins and fats. Blood
Coffee	High BOD and suspended solids
Cane sugar	High BOD mainly of carbonaceous nature
Breweries and distilleries	High dissolved organic solids mainly of fermemeted starches or their products and organic nitrogen
Textiles	High BOD, colours, high pH. high suspended solids
Leather	High BOD and total solids, sulphides, hardness
Pulp and paper	High suspended, colloidal and dissolved solids, organic matter mainly of cellulose, high or low pH, colours
Natural rubber	High BOD and odours, high suspended solids, chlorides
Starch	High BOD mainly due to starch and related material
Oil refineries	High BOD, odour, sulphur compounds, phenol
Candle manufacturing	Organic fatty acids

Rural run-off originates from the sparsely populated and from natural vegetated areas. The run-off contains variable quantities of organic matter derived mainly from the dead plant matter of the annual and deciduous vegatation. The run-off passing through the agricultural areas maso pick-up substantial quantities of organic manure and dried parts of the crop plants.

The rural run-off in general, is of diffuse nature with no point source.

Organic matter produced indigenously within the water body is the result of mainly primary and secondary production. The advanced stages of eutrophication often show the symptoms of organic pollution due to accumulation of excessive dead organic matter which exhausts the oxygen resources of water bodies and convert them into hypereutrophic or dystrophic ones.

The growing plants and animals also secrete and excrete a variety of trace organics in natural waters. The algal and fungal growths produce several organic compounds which impart characteristic tastes and odours to the waters. Some of these important exudates are sulphur containing dimethyl sulphide, methyl mercaptan, isopropyl mercaptans and butyl mercaptans. Actinomycetes exude a compound called geosmin. The decomposition of plant materials leads to the release of several coloured substances, mostly refractive in nature, the exact nature of many of them is still unknown except for tannins. Most of these compounds are polymeric and phenolic in character with high molecular weights, and are commonly called as humic substances. The excreta of aquatic animals also contains several refractory organic compounds. Though these trace organics cause mainly aesthatic problems, it has been found that some of them can produce chloroform or other halomethanes in water (Sawyer and McCarty, 1978).

Chapter-6

OIL POLLUTION

The world seems to have been blissfully ignorant about oil pollution. Oil pollution has been an almost inevitable consequence of the dependence of a rapidly growing population on oil based technology. The use of natural resources such as oil on a grand scale, without losses, has been almost impossible. The extent of such losses, intentional or accidental has been steadily increasing and has been becoming a great cause of concern.

The crude oil gets spilled, polluting the land and water. The spilling can occur from faults in the cap-rock and overlying strata or cracks in the rock strate crossing through the unprotected portion of the soil well. Spills also occur from oil shipment or from underground pipelines. Most crude oil buying countries have their own refineries, so that huge quantities of oil is transported in oil tankers, which sail through the sea and oceans around the world. Spills occur from the ships carrying crude oil to different destinations or from refining units. It has been estimated that the total oil influx into the oceans was between 5-10 million tonnes annually.

CHARACTERISTICS OF OIL

Oil is a natural substance, extracted usually from the upper strata of the earth, and commonly called petroleum. Oily substances are also derived from coal, animals, vegetation and can be made synthetically. Marked compositional differences exist, between hydrocarbons from living organisms and from fossil fuels. There is no comprehensive definition of oil, but the oily substances have several properties in common, like lighter than water, immiscible with water, flammable, spread on water forming slicks, and generally liquid at room temperature.

Crude petroleum is a mixture of several hydrocarbons, including substances bearing nitrogen, sulphur, and oxygen. Trace metals in the complexed or uncomplexed forms are also usually present in petroleum. The substances present in petroleum can be classified variously depending upon their chemical structure, physical properties or their use. Chemically they fall in two broad categories of aliphatic and aromatic compounds (Figure 23). Aliphatic compounds are with saturated or unsaturated straight or branched chains, while aromatic compounds are unsaturated having ring structures. Molecular weights of the individual components range from 16 (methane) to over 20,000 of some heavier compounds. Physically, the petroleum has light oil fractions with high volatility such as gasoline and kerosene, and heavy oil fraction like tars and residual fuel oils.

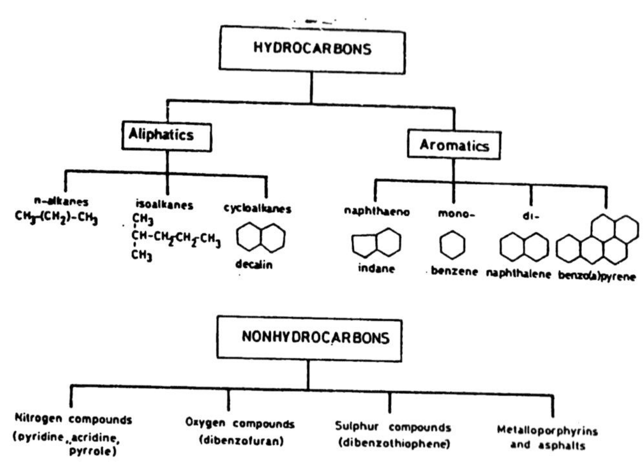

Figure 23. Classification of substances present in petroleum on the basis of their chemical structure.

The substances, according to their use, can be categorised as fuels, lubricants, coatings, cleaners, solvents, cutting and rolling fluids and hydraulic fluids.

SOURCES OF PETROLEUM OILS IN THE ENVIRONMENT

The global production of petroleum oil was more than 3100 x 10^6 tonnes out of which 1.9 x 10^6 to 11.4 x 10^6 tonnes reached the oceans through various routes (Connell, 1981).

The magnitude of oil discharged into the oceans from various sources have been shown in Figure 24. The maximum oil, which reaches the oceans, was from the oceanic transport of oil by. tankers. Most of the sea routes are along the continental shelf (Figure 25) which effects the productive coastal waters. A large quantity of oil upto 1 x 10^6 tonnes can reach the oceans by ballast water. The carriage of oil is a one way operation, and tankers on their return journey have to come without any cargo. However they fill water in the empty oil tanks, called ballast water, in order to maintain the balance and steadiness for smooth sailing. A large quantity of oil still remains in the tankers even after complete emptying, which mixes with the ballast water and reaches finally to sea when the tanker is refilled with oil after cleaning the tanks.

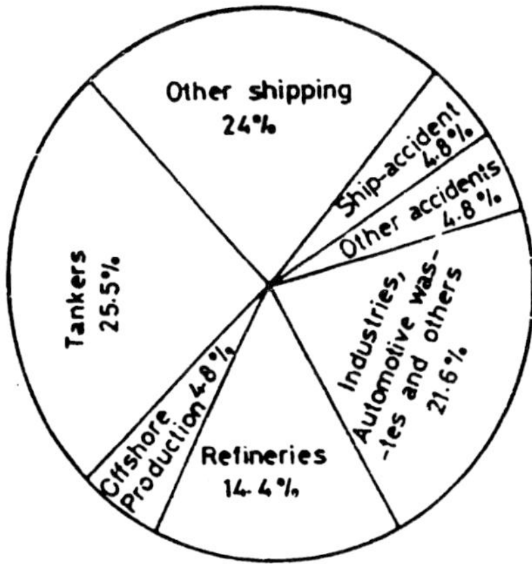

Fig. 24. Magnitude of oil in per cent, discharged into oceans by various sources.

Fig. 25. Major sea routes of oil transport.

Accidental colloison of tankers is another great source of oil in the sea that may bring about most severe ecological damage. A significant quantity of oil is also released from various industrial operations. Petroleum refinery is the principle source of oil in all the components of the environment. The use of oily materials such as lubricating oils and a number of other industrial processes, release a substantial quantity of oils in waste waters, some important industries which discharge large quantities of oils and metal industries, coke units, meat and fish processing units, wollen textiles, and paint industry. Municipal waste waters also have considerable oil derived from cleaning, automobile washings, sewage treatment units, and the degraded asphalt from roads. The total quantity of oil is, however, comparatively small from these sources.

The offshore oil production accounts nearly 5 percent of all the sources responsible for oil entry into the marine regions. Commercial and naval shore facilities are often a source of oil leakages, spills and deliberate discharge. Oil terminals, where tankers empty their oil, and the refuelling facilities for ships and boats, are also a source of oil into the sea. A much larger quantity may, however, come from natural marine seeps which range from 0.2×10^6 to 7×10^6 tonnes per year.

The global petroleum operations and uses release oily materials into all the three components of the ecosphere, that is, hydrosphere (oceans), lithosphere and atmosphere. The total global anthropogenic discharges of petroleum into the environment have been estimated to be in the order of 64.5×10^6 to 89.5×10^6 tonnes which account for nearly 2.3 to 3.2 percent of the total production (Connell, 1981).

The global dynamics of the oil with the sources and mechanisms of the transfer from one component to another have been shown in Figure 26.

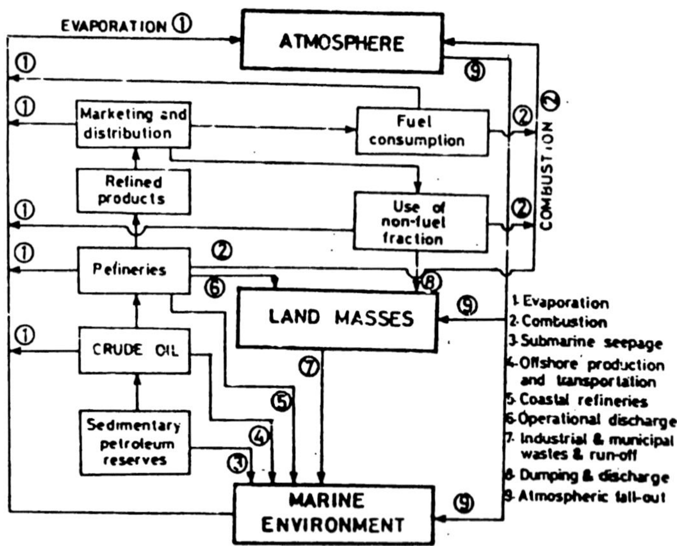

Fig. 26. Global dynamics of oil

(a) Oil seepages

Commercial exploitation of petroleum began in 1859 when E.L.Drake drilled firt oil well in Pennsylvania, United States. Early man, however, was aware about the petroleum and some of its uses through the natural seepage that occurred owing to the cracks in the cap-rock and overlying strata. There are several references to these natural seepages in the Old Testament, as slime pits which often made the combats difficult. Centuries ago, the Dead Sea was named lake Asphaltites, due to seepage of oil in the sea. Even today, floating chunks of asphalt can be seen on dense salt water of the lake. Marce Polo reported numerous oil springs from which hundreds shipload could be taken at one time, in Baku area on the shore of Caspean sea. The natural gas seepage, which often caught fire and burned continuously, were known as "eternal fires" in the early literature.

(b) Leakages from oil wells

In addition to oil seepage, leakages of oil of varying magnitude occur during drilling and under normal operations, along a fault zone crossing an offshore well (Figure 27). Such leakages are often easily managed, but some go beyond control and can lead to major accidents, sometimes leading to blowouts. One of the earliest major accident occurred at the Santa Barbara oil wells in California in 1969. The blowout in the well occurred due to a fault in the seabed starting from the point where the well had no protective steel casing. The massive oil slick of 3150 Kl oil leaked out for 300 days and spread to the shore line covering more than 250 square kilometers of the ocean surface. The well which was involved in the blowout had steel casing only upto a depth of 73 meters. Below this the well passed through weak rock of claystone and sandstone. The blowout occurred due to fracture of this rock under gas pressure. Another major accident involving an oil well occurred in Mexico in 1979. More than 500,000 tonnes of crude oil were discharged into the sea between June and November from this well.

OFF SHORE WELL

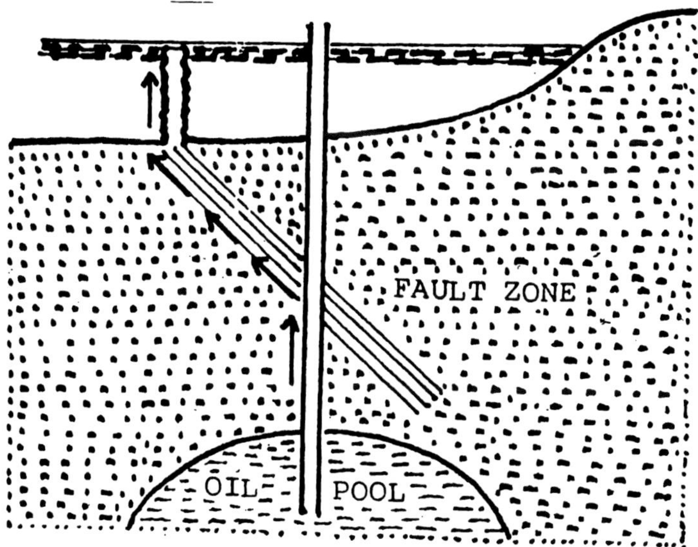

Figure 27. Crude petroleum seepage along a fault zone crossing an offshore well.

(c) Tanker spills

A major source of oil pollution are oil tankers that sail around the world. These tankers spill oil into the oceans regularly due to hose failures, pumping errors, tank and bilge cleaning operations and leakages of lubricating oil from propeller shaft bearing:

In an insidious method that escapes public notice, oil waste is deliberately discharged into the sea. When tankers reach their destination, the oil is taken out, the empty tankers are filled with sea water, to avoid floating toc high on their return journey, which will be hazardous as well as uncomfortable for the crew. This dirty water is pumped out when the tanker takes in fresh cargo. It is this practice, which leads to the presence of tar-balls found on our beaches. This widespread practice of using sea-water as ballast for empty tankers contributes an estimated 3 million tonnes of oil to the oceans annually.

The dumping of bilge contents by ships other than tankers contributes an estimated 500,000 tonnes per year.

In-port oil losses takes place during loading and unloading procedures and also as a result of collisions in port. The estimated annual contribution to oil pollution of the sea has been 1 million tonne.

Oil tanker accidents on the high seas or near the shore certainly add to the problem and receive great amount of publicity, The first oil spill occurred when the tanker *Torrey Canyon* accident occurred in the English Channel on March 18, 1967, spilling 95,000 tonnes of its cargo of 1,17,000 tonnes of crude oil into the sea.

Oil tanker disaster caused pollution problem in Antarctica. The Argentine ship "Bahia paraiso" wrecked near the Antarctica shores in Januaty 1989, spreading the oil in a radius of several kilometers. Major oil tanker disaster include an accident involving a Greek tanker "Exxon Valdez" which hit the Brenton Reef in Rodhe island in June 1989 and spilled nearly 45 million litres of oil into the Alaskan sea, that spread over an area of nearly 1200 Km^2. Large quantities of oil drifted to the tourist beaches, spoiling them very badly. In the same month, an Uruguayan oil tanker also ran aground in Dalaware river about 32 km south of Philadelphia leaving spill of about six million litres of oil.

In one of the worst oil spills in the Arabian sea, affecting the Indian coasts, a Maltese oil tanker "Puppy" collided with a container vessel "Wrold Quince" about 550 km away from Mumbai on June 28, 1989. It spilled nearly 5000 tonnes of oil into the ocean. The oil slicks moved even to the beaches of India.

Since most of the oil is produced in Gulf countries, the main routes of oil transportation run along Arabian Sea. One of these is through Mozambique Channel around South Africa to Western hemisphere. The other route is around Sri Lanka across southern Bay of Bengal, reaching Malacca Strait to Far-east. Indian ocean is thus the most vulnerable to oil pollution.

Estimates of widespread oil pollution are based on collection of oil-tar lumps over wide reaches of the oceans. The concentrations of 0.02 - 0.6 mg/m^2 have been reported in North Atlantic ocean, 0.12 - 13 mg/m^2 in North-west Pacific ocean, 0.3 - 112 mg/m^2 in Arabian Sea and 0-70 mg/m^2 in Bay of Bengal. It has been estimated that at any time, the amount of oil-tar floating on surface layers of Arabian Sea would be about 3700 tonnes while along the Southern Bay of Bengal it would be 1100 tonnes.

One of the worst oil spills occurred in Januray 1991 in the Persian Gulf, during the Gulf-war between Iraq and the multinational forces led by th United States. Iraq allegedly pumped out 10,000 tonnes of oil per day in the sea, dumping several millions barrels of oil in the Persian Gulf. Iraq, however, maintained that oil spills had been caused by bombing of two of its oil tankers by the multinational forces. The oil slick extended to length of 140 kilometers and a width of 40 kilometers.

Besides these sources of oil pollution, oil is also released into the sea from land in the form of refinery wastes, unburned fuels, lubricants or even from the atmosphere as fall out of petroleum particles and products injected by automobile exhausts and industrial activities (Aggarwal et al, 1993).

THE FATE OF SPILLED OIL

Oil and water do not mix. Therefore, spilled oil just floats until it is washed ashore. This however, is not the case. One cubic metre of oil spreads within 10 minutes to a patch of 48 metre diameter and after 100 minutes the oil film becomes only 0.1 mm thick, spread over 100 metre. During the time it has been spreading, many changes take place in the oil. All oils, regardless of types, have volatile components that evaporate readily. Within a few days 25 percent of the volume of a typical oil spill is lost through evaporation. The remaining oil has been subjected to emulsification processes which, in a sense, do cause oil and water to mix.

Two types of emulsions have been formed. Oil-in-water emulsions result when the droplets of oil get dispersed in water and are stabilized by chemical interaction with the water at their surfaces. This process occurs especially well in turbulent seas. The resulting oil droplets do not get dispersed on top of the water, but rather throughout the water. Some oil droplets, especially those weighted down by mineral particles, even reach the bottom.

Water-in-oil emulsions form when water droplets get enclosed in sheaths of oil. The emulsions gets stabilised by interaction between the enclosed water droplets and various resinuous and asphaltic materials found in most crude oil. This type of emulsion shows up a floating, ticky, viscous mass that on occasion, contains enough water to cause the total volume to be greater than that of the original oil.

Most oil that survives the emulsification process gets degraded by spontaneous photo-oxidation and oxidation by microorganisms. The micro-organisms have been the most powerful source of oil decomposition found in the sea.

By the end of three months at sea only about 15 percent of the original volume of an oil spill remains. This is in the form of black, tarry lumps of a dense asphaltic substance, and it has been those lumps that frequently reach the beaches.

When massive spills take place close to shore, sufficient time is not available for the process described above to affect the total amount of oil involved. Under such circumstances a thick ticky film of oil gets deposited on any solid that comes in the contact with the spill.

The petroleum pollutants in the oceans may occur at any concentration between bulk oil and oil in very small quantities, which can remain in the following forms:

1. Floating material
2. Emulsion dispersed in seawater

3. Dissolved in seawater
4. Adsorbed on sediments
5. Locked in marine organisms

The freshly spilled oil in the seawater, after inter-acting with the prevailing physico-chemical conditions and biological factors, undergoes compositional changes. From the surface of the sea, it moves to other components in modified and unmodified form by a variety of mechanisms (Figure 28). The light fraction of the oil in readily evaporated to the atmosphere by wind spray and bursting of water bubble. Photo-oxidation of the surface oil forms a variety of toxic and nontoxic materials such as free radicals, carboxylic acids, esters, oxygenated aromatics, and carbonyl compounds (Hansen, 1977). Many of these substances ultimately find their way into the atmosphere, while the rest mix with water. Much of atmospheric oil, falls-out again on the water surface.

With the increae roughness of the surface, some oil droplets form the oil-in-water emulsion (oil droplets in water), which ultimately sinks to the bottom incorporating silt and suspended particles into it.

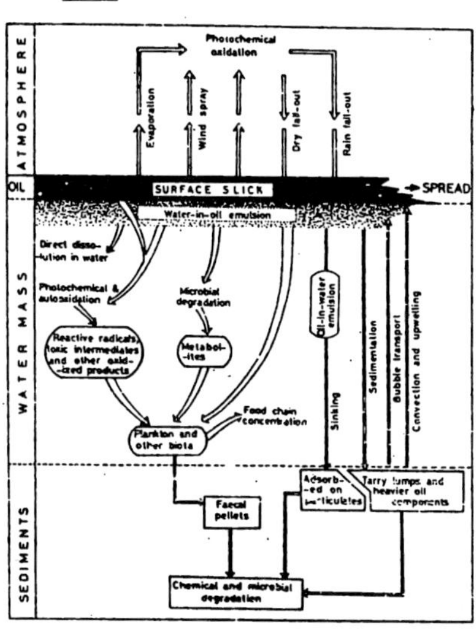

Fig. 28. Fate and movement of oil in marine environments.

The residual oil spreads over the water surface forming a thin layer, and doing so, it is stirred with water to form a water-in-oil emulsion (water droplets in oil), which still behaves like an oil, yet it may contain nearly 80 percent of water. Several bacterial species and a few fungi in water and sediments metabolize a number of hydrocarbons of crude petroleum. The oxydation is usually species dependent, and is influenced by the factors like salinity, water temperature, nutrients, extent of dispersion in water, chemical composition of oil and abundance of micro-organisms. The microbial degradation of oil results in further formation of a variety of products. These newly formed metabolites, together with some other oil components, are later widely consumed by. other ambient concentrations of these oil fractions.

Most oil fractions are normally lipophilic in nature with a very low solubility in water, and tend to accumulate in fatty tissues of the organisms. The absorption mode of the oil may be through respiratory surfaces, the gastrointestinal tract and external surfaces. Concentration factors for certain oil fractions in a clam (*Rangea cuneata*) have been shown in Table 87. Some persistent petroleum hydrocarbons have also shown their biomagnification along the food chains (Teal, 1977). Miller and Connell (1980) reported high concentration of oils in some sea birds as a result of biomagnification (Figure 29).

Table 87. Concentration factors of certain oil fractions in the clam *(Rangia cuneata)* (Connel & Miller, 1984).

Oil component	*Concentration Factor*
Naphthalene	6.1
Phenanthrene	32.0
Chrysene	8.2
Banzo[a]pyrene	8.7

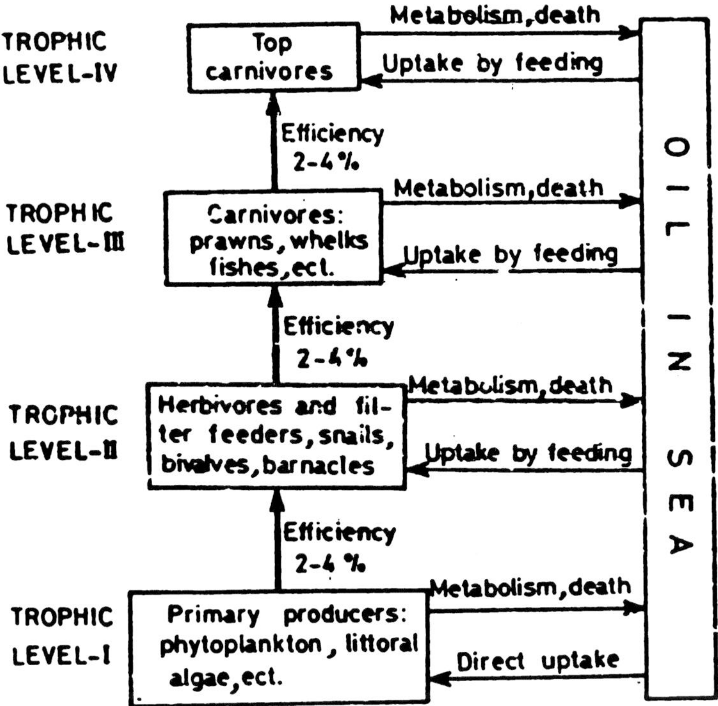

Fig. 29. Marine food chain showing the concentration of oils.

PHYSICAL AND BIOLOGICAL EFFECTS OF OIL POLLUTION

The problem of oil pollution may be thought in terms of short-term and long-term effects. The short-term effects have been immediately obvious and have been the ones that have received the most publicity. The long-term effects have been only slowly becoming apparent and have been currenty the subject of much study. The short-term effects fall into two categories, those caused by coating and asphyxiation, and those that result from oil toxicity.

1. Reduction of light transmission

Measurements have regarded that ambient light intensity 2 meters below an oil slick was 90 percent lower than at the same depth in clear water. Photosynthesis by marine plant life may get hampered by extended periods of such conditions.

2. Dissolved oxygen depletion

Oil film are able to retard the rate of oxygen uptake by water. Measurements have shown a significantly lower dissolved oxygen level in water beneath a slick than in clear water.

3. Damage to marine life

Floating layer of tar result in considerable damage to the marine life. The oily coat breaks down natural oils and waxes that insulate and protect the swimming and diving birds from loss of body heat. Their feathers get matted, reducing their buoyancy in water and preventing their flight. The odorous materials in the oil tar cause a lot of confusion in migratory activities of marine life. Lowered dissolved oxygen level of marine water leads to death of marine life. Low boiling point components of oil are more toxic and their greater solubility enhances their uptake by the organisms. The polycyclic aromatic hydrocarbons of oil may have mutagenic and carcinogenic effects. High boiling point hydrocarbon fraction of crude oil interfere with responses to chemical stimuli, such as sex attractants. Large scale oil spills alone spell doom for marine life. Because of transportation of large volume of crude oil through the Indian ocean, many areas are being affected. The worst affected are coral reefs and beaches. Significant damage has been observed from atolls of Lakshadweep, coral reefs of Andaman and from Seychelles group of islands.

Oils affect the organisms in a number of ways depending upon the characteristics of the oil fractions and their concentration in water. The effects of oil, in a broader sense, can range from purely mechanical to various toxic effects. Miller (1982) has pointed out that the responses of organisms to various concentrations of oils range from simple bioaccumulation to deaths of biota (Table 88). Many studies have been carried out by various workers (Baker, 1970; Rice et al, 1977; Moore and Dwyer, 1974, ESAMP, 1977; Michael, 1977; Lenihan and Fletcher, 1977;

Neff, 1979; Connel and Miller, 1981 a, b; and Miller,1982) to understand the effect of oil on various organisms.

Table 88. The responses of marine organisms to various concentration of soluble aromatic hydrocarbons (After Miller, 1982)

Concentration (ppb)	Response
0-10	Bioaccunmlation
10-100	Effect on behaviour
100-1000	Effect on growth and reproduction
1000-10000	Lethal to larval and juvenile forms
> 10000	Lethal to adults

The bioassay tests carried out by different workers reveal that lighter oils and water soluble fractions of fuel oils are comparatively more toxic than heavy oils and water soluble fractions of crude oil. In general fish and crustaceans, particularly decapods, are more sensitive than others. Cold temperature biota is more sensitive to oil than their counterparts warm temperature species. Intertidal species are more tolerant to oils than others. Differential sensitivity has also been reported with different stages in the life history such as eggs, larvae and adults, with the oil more toxic to larval and juvenile forms. Miller (1982) studied 96 hour LC_{50} of some soluble aromatic fractions of crude and refined fuel oils for different classes of organisms (Table 89).

Table 89. LC_{50} values of aromatic water soluble fractions of crude and refined fuels for certain groups of marine biota (After Miller, 1982).

Group of organisms	Crude oils	Refined fuel oil
Finfish	l-> 10	l-> 10
Crustaceans	l-> 10	0.1-> 10
Bivalves	l-> 10	0.5->5
Gastropods	l-> 10	l->5
Other benthic invertebrates	> 5-10*	> 1*
Larvae (all species)	0.10-5	*
Juveniles	5-> 10	1-10

* Inadequate data

4. Smothering

The effects of oil on plant life along shorelines are becoming a matter of concern. Intertidal algae and lichens are known to have been killed by smothering coats of oil which gets washed ashore in some areas.

The toxic effect of oil is because of the many compounds that make up this very complex mixture. These compounds have been found to vary widely in their molecular weights and structure. Until recently, low boiling saturated hydrocarbons were regarded to be harmless to marine life. It has now been demonstrated that these substances, in low concentrations, cause anaesthesia and necrosis in a wide variety of lower animals. Cell damage and death result in animals exposed to high concentrations. The larvae and other young forms of marine life may get especially affected.

Low boiling aromatic hydrocarbons are abundant in oil and represent its most dangerous fraction. Benzene, toluene and xylene occur in this fraction. These substances have been known as acute poison for man and other living organisms. Naphthalene and phenanthrene are also present in oil. These latter compounds are even more toxic to fish than the former three compounds.

The aromatic compounds are more water soluble than the saturated hydrocarbons. The aromatic compounds kill marine organims by direct contact (with oil) or through contact with dilute solution of oil in water. The adverse effect of these compounds diminishes with time because of the volatile fraction which evaporates away. The long-term (low concentration) effects of oil components on living organisms is not as apparent as short-term effects.

The fate of organic compounds which enter the marine food chain indicate that hydrocarbons, once incorporated into a particular organism, become stabilized, regardless of their structure. There is evidence that these compounds can mass through many members of the marine food chain without being altered. Such compounds are thus subject to food-chain amplification analogous to the situation found for heavy metals and pesticides. This presents the possibility that the poisons might accumulate in the sea-food harvested for human consumption. This is more serious because we know that high-boiling aromatic hydrocarbons are carcinogenic (Tyagi and Mehra, 1990).

Two million tonnes of oil are ferried across the Arabian Sea annually. In the recent past we have had several spills from tanker accidents, the latest one at 0/6 Nicobars in 1993, incurring a loss of an estimated 15 crores in terms of loss of fisheries alone. Yet sporadic accidents are small threat to a politically well-connected Reliance Oil Company's proposals in the Gulf of Kutch (one of our three marine biosphere reserves). Their preposterous attempt to denotify the reserve is not only making a mockery of any positive conservation effort but a laying waste some of our best marine areas, harbouring 1000-odd species from the rich fisheries of the Gujarat coast. This will also destroy what limited scope of pearl culture we have in this area. The pollution hazard is not only from the laying of pipelines over coral beds or hypothetical spills (Paul, 1995).

OIL POLLUTION CONTROL

Some of the common methods to control the oil pollution from marine and other water bodies have been briefly discussed below:

(a) Mechanical containment

The oil spills spread in the form of slicks which can be contained by means of some mechanical barriers. One such method is the use of floating containment barriers or booms. These are made up of relatively compatible plastic materials, such as polyethylene, polyurethane, polyvinyl chloride, nylon, neoprene, epoxies, and polupropylene, and contain inflated sacs at regular intervals for flotation (Figure 30). Any floating or spreading oil can be contained in them from were it can be removed by some other means.

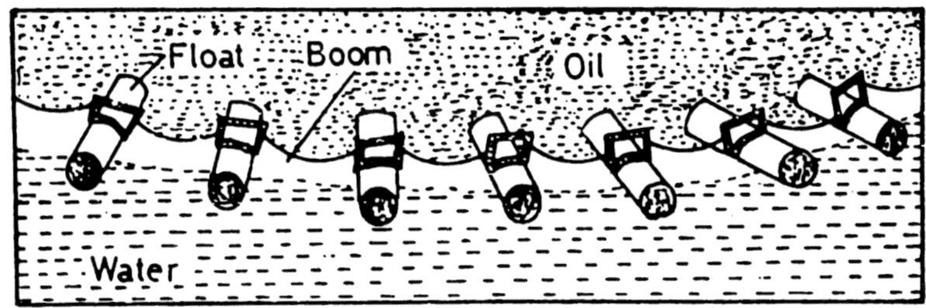

Fig. 30. Floating containment barrier (boom).

(b) Mechanical recovery

After the containment, the oil can be removed by a number of means such as by use of weirs, suction devices or by lifting surfaces. Weirs (Figure 31) are normally more efficient where the oil is comparatively deep, in thin layers of oil, a substantial quantity of water often comes with oil. The lifting surfaces are oleophilic and hydrophobic which move through the oil. While moving some quantity of oil gets attached to them, which can be extracted and pumped after withdrawal of these surfaces from the oil. These lifting surfaces may be in the form of vertical discs and vans which rotate through the oil and are fixed on suitable inflatable carriers.

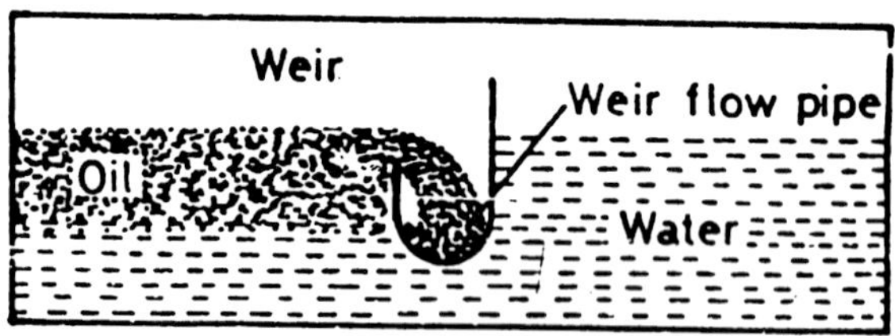

Fig. 31. Removal of oil by use of weirs.

(c) Combustion

Combustion was one of the earliest methods used to rid water of oil pollution. It has been an attractive and inexpensive method for disposing of large quantities of oil. However, it is inadvisable in situations where the oil spill is close to the shore or to structures that could be damaged.

Generally, burning is effective only if it can begin immediately after a spill. The lighter fractions of oil evaporate immediately after a spill, and the remaining oil is having a higher ignition temperature rand can be difficult to ignite. This factor, coupled with the ability of water to conduct heat away from the burning oil, has caused some attempts to be unsuccessful. Research is now being aimed at developing "aids to combustion" in the form of wicks, catalysts and supplementary fuels. Further developments in burning techniques may give rise to methds very useful on large spills in the open sea. One drawback has been that water pollution has been converted to air pollution. The residue left after burning has been a black solid which is easier to collect than the liquid oil.

(d) Sorbents

Sorbents can remove small quantities of oil. Oil gets soaked up by an absorbent and clings to the surface of particles of an absorbent. Straw and saw dust are used as sorbents. After becoming saturated with oil, they might be removed and disposed of. Shredded polyurethane foam, an absorbent, is used and has the advantage that, when it gets saturated, it can be removed and the oil can be squeezed out. The foam can then be reused. The recovered liquid is 80 percent oil, which can be recovered for use. Sinking agents - sand, brickdust, cement, silicone mixture are used to sink oil. Pulverised chalk and sodium stearate, or even sand plus ash coated with a silicon compound to make oil sink. However, this smothers and poisons the organisms living at the sea bottom.

(e) Emulsification

Emulsification is yet another way of getting rid of the oil slick. Many materials can emulsify oil in water. These materials are known by a variety of names, such as emulsifiers, detergents, degreasers and dispersants. The primary components of a dispersant have been a surfactant, a solvent, and stabilizers. The surfactant has an affinity for both oil and water and allows the two to mix. Solvents enable the surfactant to penetrate into the oil slick. The solvent usually makes up the bulk of the dispersant composition. Stabilizers are able to fix the emusion once it is formed. The net result of the dispersant is to lower the surface tension of the oil to the point where it will break up and disperse in water in the form of tiny droplets.

The action of dispersants is able to increase the surface area of an oil slick and distribute the oil droplets into a large volume of water. The chemical dispersants themselves are not able to destroy any oil. Psychologically, it is a case of "out of sight, out of mind". The authorities may claim that the slick has disappeared into smaller droplets, but the public do not realise that oil broken into smaller droplets will mix readily with sea-water, thereby enhancing the poisnous effect of the oil. Eventually, the oil will come out of its emulsified state and affect the areas again.

Even in crises management methods, different interests come into play. Thus, hoteliers, beach resort owners and the tourism industry try their best and persuade the authorities to prevent the oil from being wahsed ashore on the beaches, where their patrons swim or sunbathe. On the otherhand, fisherman cry foul if detergents or sinking agents are used, which ruin their catches and affect their livelihood.

The toxicity of fuel oil number one is 20 ppm for fish and 0.4 - 0.6 ppm for other marine life. The rate for degradation by natural means vary from 36-360 micrograms per square metre a year. The isoprenoids, alicyclic and aromatic compounds of crude oil can be detected even two years after an oil spill. Even one and half year after a 600 tonne of oil spill into the sea, the mud at the bottom contained 70 mg hydrocarbon per kg (of mud) which was still poisnous to marine life.

(f) Bacterial degradation

Bacetrial degradation of oil is a more eco-friendly way. Bacteria are used to "eat up" the oil. Many bactria eat components of oil, but no single bactria can decompose all the components. Here genetic engineering has come to our help by combining the abilities of several different bacteria to produce a "supergerm" which can break up all the components of crude oil, This is much quicker than the natural biodegradation of oil.

The best way out of these predicament will be to contain the oil before it spreads out by using floating booms, These should be 60-120 cm above and below the sea surface. The oil is then scooped up or sucked in by pumps and the water separated. However, this method works only when the layer of oil in thick. While it is feasible in harbours and calm areas, it does not work in turbulent whether or in rough seas with high waves (Chhapgar, 1993).

Chapter-7

SEWAGE POLLUTION

Sewage is the most general term used to define all kinds of liquid wastes produced in a community. It is the liquid form residential and industrial establishments of a town. It is made up of spent water from bath rooms, lavatory basins, kitchen sinks and the discharge from urinals of homes, offices and institutions, human and animal faeces. The sewage is extremely foul on account of the human excreta (or night soil) and urine contained in it, which putrify and give out offensive smell.

PHYSICO-CHEMICAL NATURE OP SEWAGE

The raw sewage has only about 0.1 percent solids and 99.9 percent water. The organic and inorganic constituents in the solid fraction are in the ratio of 70:30. The inorganic constituents include grit, salts and metals in varying proportions while organic substances are proteins, carbohydrates and fats. An analysis of typical Indian, American and English sewage (Table 90) and some cities of India (Table 91) indicate that Indian sewage is comparatively richer in various parameters than those of American and English.

The characteristics of sewage of Kholapur (Table 92), Chembur (Table 93) and Ujjain (Table 94) indicate that the pH ranged from 6.7 to 7.65, under a safe range as stated by Ellis (1937), Roberts et al (1940), and Klein (1973). These authors have clearly mentioned that the pH value between 6.7 to 8.4 was suitable, while a value below 5.0 and more than 8.8 was detrimental. A significantly higher amount of dissolved solids have been reported, buv the quantity was more than suspended

Table 90. Analysis of typical Indian, American and English sewage results expressed in parts per million (Kshirsagar, 1968).

Constituent	Ahmedabad	Poona	Maduai	Nagpur	Hyderabad	Lucknow	San Antonio Texas	Cleaveland Ohio	Durham N.C.	Modgen	Burming-gam (U.K.)	Leeds High Land
Total Solids	1748.0	...	1740	2200	1708	2240	1071	678	712	...	...	888
Suspended Solids	148.0	1098.0	420	800	985	1710	241	184	227	307	604	223
Dissolved Solids	1600.0	...	1320	1400	723	530	830	494	485	...	..	665
Ammoniacal Nitrogen	225	57.6	...	34.0	24.8	62.4	16.1	14.0	15.1	40.8	23.1	...
Albuminoid Nitrogen	4.8	57.8	4.0	430	26.1	23.0				9.5	7.5	167
Bio-chemical Oxygen demand at 20°C	133.0	234.0	480.0	232.0	338.6		125	163	333	357		
Oxygen absorbed from KM_nO_4 in 4hr.	67.0	127.0	88.2	161.0	...	...			161		211	
Chlorides as Cl	300.0		280.0	120.0	104.5	60.0		67.0		114	107	117
Phosphates	...	...	...	...	...	16						
pH	7.5-8	7.0	7.5	7.2	7.3	7.5	...	...	...	...	...	...

Table 91. Chemical composition of sewage of some Indian cities (Kant, 1987).

Parameters	Jaipur 1	Agra 2	Bangalore 3	Madras 4	Nagpur 5	Kanpur 6
pH	7.3-8.2	7.1-8.2	7.1	7.1-8.2	7.2	7.0
Total diss. solids	832.0-2368.0	530.0-828.0	–	750.0-1400.0	1000.0	900.0
Total susp. solids	321.0-601.0	725.0-812.0	–	119.9-730.6	200.0	600.0
Sus. Org. matter	272.0-403.0	–	–	–	–	–
COD	–	12.0-42.0	–	624.0-1120.0	–	–
BOD	–	112.0-162.2	–	–	–	–
Chlorides	198.0-430.0	134.0-270.0	138.0	170.0-280.0	30.0	85.0
Calcium	70.0-128.0	11.0-27.0	–	96.0-172.0	–	–
Magnesium	37.0-91.0	13.0-26.2	–	70.0-76.0	–	–
Ammonia	10.0-46.0	3.8-11.5	39.0	–	60.0*	50.0*
Nitrate-N	Tr-1.6	0.06-0.21	–	Tr	–	–
Phosphate-P	0.9-7.9	–	6.5	–	20.9	15.0
Total hardness	–	346.0-452.0	–	–	–	–

1. Goel *et al.* (1980); 2. Sharma *et al.* (1981). 3. Shridhar & Pillai (1973). 4. Raman *et al.* (1973). 5. Thakur *et al.* (1977).

Tr = Traces. * = Total nitrogen All the values are in mg/l except pH.

Table 92. Characteristics of sewage of Kolhapur (Trivedi, 1982b).

Particulars (mg/l)	Before treatment	After treatment
pH	6.7 - 7.65	6.9 - 7.25
Total solids	600 - 870	450 - 770
Dissolved solids	300 - 550	300 - 550
Suspended solids	240 - 340	150 - 220
BOD	70 - 387	80 - 220
COD	–	–
Chlorides	65 - 84	20 - 81
Sulphates	10 - 26	6 - 28
Oil and grease	24 - 86	20 - 40

Table 93. Quality of raw sewage, treated water and water from municipal main at Chembur (Bhoota, 1975).

Particulars (mg/1)	Raw sewage	Treated water	Municipal water
pH	7.15 - 7.65	7.1	7.8
Total hardness	120 - 162	140	39
Alkalinity	125 - 200	135	40
Chlorides	96-130	60	18
Dissolved solids	550 - 640	300	90
Suspended solids	70 - 100	nil	10
Turbidity	–	3	6
B O D at 20°C	200 - 250	1.0	1.2
COD	250 - 350	3.5	1.7

solids, which might be because of the gradual sedimentation in the bottom of the sewage drains (Motwani et al, 1956? Verma and Shukla, 1969). The chlorine concentration in both the cases was high, and they belonged to strong category of Klein (1957) and suggested their pollution by organic matter.

Table 94. Composition of sewage waste water from sewage farm, Ujjain (Shrivastava and Singh, 1990).

Parameter	Values
pH	7.4
Conductivity (Millimhos/cm)	2.2
Nitrogen (µg/ml)	54.0
Sodium	117.0
Potassium (ppm)	63.0
Cadmium (ppm)	0.05
Zinc (ppm)	4.00
Chloride (mg/1)	412.0
Total dissolved solids (mg/l)	1500.0

The measurement of the quantity of nitrogen has been considered as highly significant to assess the pollution condition as the domestic sewage has been one of the important sources for the increase in the nitrogen concentration (Sylvester, 1961; and Adoni, 1961). The organic pollutants present in water remained in solution and mainly consisted of protein, nitrogen and ammonia. Ammonical nitrogen level was high in comparison to nitrate nitrogen in water. The organic matter consisted primarily of carbohydrates, proteins and miscellaneous fats and oils. The specific classes of organic compounds found in sewage include amino acids, fatty acids, amides and many others. Dissolved salts in the form of ions such as sodium, potassium, calcium, manganese, ammonia, chlorides, nitrates, bicarbonates, sulphate and phosphate are the main inorganic constituents of sewage and other waste waters (WHO, 1972).

In Kolhapur (Maharashtra) about 50 to 55 mld of sewage is generated. The topography of the city is such that all the waste water can be collected at one location near collectorate by open line channels and by closed pipes, However, these open drains are not maintained water light, this leads to occasional over flow, spillage, leakage, percolation in the sub-soil strata. In many cases the channels do not meet the designed task, but go to two open drains in the city. These drains meet the river Panchganga. Out of the total population 40 percent is not served with any sewerage system. The waste water collected by the Municipal Corporation is given primary treatment and afterwards diverted to the agriculture fields for irrigation without any significant dilution. The water not utilised for irrigation is channelized to the river Panchganga. The site of disposal is upstream of the drinking water uptake site.

SEWAGE FARMING AND ITS EFFECTS

Sewage farming is practiced in India by using mostly raw sewage and some times partially treated sewage. There are about 132 sewage farms in the country covering approximately 1200 hectares and utilizing 1 million cubic meters of sewage per day. In cities like Bangalore, Calcutta, Madurai, Madras and Mysore etc. sewage farming is practiced (Sastry, 1986).

The world's largest combined waste water disposal and fish production system is in India, Waste water and drainage water from East Calcutta are stored and treated in over 40 square kilometer of lagoons where fish production has become a highly skilled operation (Edwards, 1985). Some 24000 people are employed, full or part-time, in this country. The output provides 10-20 percent of the fish requirements of 10 million urban dwellers.

In selecting the crops for sewage farms, the major consideration should be public health. The use of untreated sewage for irrigation is unadvisable because its handling by the farmers and eating of the produce from these farms pose serious health hazards. Contamination of crops grown on effluents, by pathogens and hookworms is considered to be one of the important hazards. Sewage farms workers do not normally use any boots or gloves on the farms and hence suffer from hookworms infestation (Sastry, 1986).

Sewage wastes, both solid and liquid, are increasingly applied to agricultural land. They contain heavy metals in varying proportions, depending on the nature of the local industry and the proportion of domestic sewage. The sewage sludge from England and Wales (Table 95) and Melbourne sewage farm (Table 96) contribute a substantial proportion of the total load of copper, chromium, lead and zinc and to a lesser extent cadmium, mercury and iron.

Table 95. Concentration of heavy metals (µg/g dry wt) in sewage-sludge from England and Wales (Berrow and Webber, 1972).

Metal	Median	Range
Cadmium	–	60-1500
Cobalt	12	2-260
Cromium	250	40-8800
Copper	800	200-8000
Iron	21000	6000-62000
Manganese	400	150-2500
Molybdenum	5	2-30
Nickel	80	20-5300
Lead	700	120-3000
Stanium	120	40-700
Zinc	3000	700-49000

Table 96. Concentration of heavy metals (µg/cm³) in liquid raw sewage and drainage water at the Melbourne sewage, Werribee, Victoria, Australia (Leeper, 1978).

Metal	Raw sewage	Drainage water
Cadmium	0.02	0.005
Cromium	0.42	0.1
Copper	0.31	0.04
Nickel	0.21	0.06
Lead	0.12	0.03
Zinc	1.1	0.08

Observations (Table 97) show that higher loading rates of sewage and sludge are also accompanied by higher organic loading (BOD) and excessive supply of nitrogen and phosphorus. Some hazardous metals are also carried along with essential nutrients. These metals include cadmium, chromium, iron, copper and zinc. A lower concentration of these metals may not become toxic for plant growth, but they get introduced in the food chain which might have toxic effects on animals and human beings.

Table 97. Average properties of sewage, sludge and tap water used for irrigation (Misra and Shukla, 1992).

Parameter	Sewage	Sludge	Tap water
pH	7.2	7.6	7.8
Bicarbonate	0.71	–	0.86
BOD	110.45	–	–
Nitrogen (ppm)	61.20	11.0	trace
Phosphorus	7.10	4.25	ND
Potassium	76.80	18.60	ND
Cadmium	0.62	5.9	ND
Lead	6.50	67.0	ND
Iron	8.75	71.79	2.01
Chromium	0.31	7.71	ND
Zinc	2.20	8.84	0.23
Manganese	5.63	15.81	ND

It has been observed that the total cadmium in sewage irrigated soils is always higher than in the non sewage irrigated soils. The total cadmium concentration in different soils, irrigated with sewage water ranged from 0.3 to 6.0 mg/kg, against 0.1 to 1.7 mg/kg in non-sewage irrigated soils (Table 98).

Table 98. Distribution of heavy metals in sewage irrigated and non-sewage irrigated soils of Sheila Dhar institute experimental farm (mg/kg) Misra and Mani, 1992).

Soil		Cd		Cr		Pb		Zn	
Treatment	Depth (cm.)	Total	DTPA extractable	Total	DTPAT extractable	Total	DTPA extractable	Total	DTPA extractable
Sewage irrigated									
	A_1 0-10	2.2	0.31	9.8	0.09	4.4	2.8	108	2.22
	A_2 10-20	6.0	016	2.6	0.08	3.6	1.8	160	1.06
	A_3 20-30	4.0	0.12	4.4	0.10	3.2	1.3	130	1.46
	A_4 30-40	1.5	0.10	2.8	0.12	1.5	0.9	125	1.86
Non-sewage irrigated									
	A_1 0-10	1.7	0.03	1.2	0.06	1.00	0.81	108	0.12
	A_2 10-20	0.62	0.12	0.8	0.01	0.92	0.34	120	0.26
	A_3 20-30	1.2	0.04	0.5	0.03	0.40	0.18	136	0.16
	A_4 30-40	0.55	0.02	0.6	0.02	0.38	0.08	116	0.12

Extractable (available) cadmium also varied in sewage irrigated and non-sewage irrigated soils. In sewage irrigated soils, the DTPA extractable Cadmium is much more than in non-sewage irrigated soil.

Irrespective of the sewage application, the extractable cadmium content decreases with depth. Average content of extractable Cadmium in normal and sewage treated soils at 0-10, 10-20, 20-30 and 30-40 cm depth was 0.32, 0.14, 0.03 and 0.01; and 1.70, 0.86, 0.50 and 0.35 mg/kg respectively. Total cadmium concentration also decreased with depth, except in Sheila Dhar Institute experimental farm soil, which is a dumped soil, hence total Cadmium concentration decreased in both types of soils after a depth of 10-20 cm.

Total Chromium concentration in all the soil samples irrigated with sewage water was much higher than non-sewage irrigated soils. It was further noticed that extractable Chromium does not vary much in sewage irrigated and non-sewage irrigated soil samples from Sheila Dhar Institute experimental farm, and K.A.P.G. College farm (Table 99), where as samples from Naini Sewage soil exhibit variable results (Table 100). It is clear that extractable Chromium is affected by the type of the soil.

Total Chromium decreased with depth (Table 95 to 96), from 9.8 to 1.2 $mg^1 kg$; and 2.4 to 0.5 mg^1 kg in 0/40 cm depth in sewage irrigated soil and in normal soil respectively. Extractable Chromium also decreased with depth in both sewage irrigated soil and in normal soil.

Most lead was found in the surface soil, indicating little if any downward movement. Total lead in almost all soil samples irrigated with sewage was much higher than non-sewage irrigated soil. Total lead content decreased with depth from 4.4 to 1.10 mg/kg; and 0.01 to 0.36 mg/kg in 0-40 cm depth in sewage irrigated soil and in normal soil respectively. Extractable lead also decreased with depth in both sewage irrigated and normal soils.

Total zinc did not show much variation in sewage irrigated and non-sewage irrigated soils. However, extractable zinc in sewage irrigated soils was always more than three times that of non-sewage irrigated soils. Zinc always decreased with increase in soil depth.

Table 99. Distribution of heavy metals in sewage irrigated and non-sewage irrigated soils of K.A.P.G.College farm (mg/kg) (Misra and Mani, 1992).

Soil		Cd		Cr		Pb		Zn	
Treatment	Depth (cm.)	Total	DTPA extrac-table	Total	DTPAT extrac-table	Total	DTPA extrac-table	Total	DTPA extrac-table
Sewage irrigated									
	A_1 0-10	1.8	1.5	4.8	0.14	3.8	2.5	105.00	6.20
	A_2 10-20	1.4	0.6	3.2	0.16	3.0	1.7	99.00	3.50
	A_3 20-30	0.9	0.3	1.8	0.09	2.6	1.3	88.00	4.60
	A_4 30-40	0.3	0.08	1.2	0.04	1.1	0.8	86.00	2.20
Non-sewage irrigated									
	A_1 0-10	0.8	0.2	2.4	0.08	1.00	0.80	104.00	2.60
	A_2 10-20	1.0	0.2	1.6	0.04	0.90	0.32	100.20	2.70
	A_3 20-30	0.1	–	0.8	0.5	0.40	0.16	68.60	2.10
	A_4 30-40	–	–	0.5	0.03	0.36	0.09	62.50	1.50

Table 100. Distribution of heavy metals in sewage irrigated and non sewage irrigated soils of Naini sewage farm (mg/kg) (Misra and Mani, 1992).

Soil		Cd		Cr		Pb		Zn	
Treatment	Depth (cm.)	Total	DTPA extractable	Total	DTPAT extractable	Total	DTPA extractable	Total	DTPA extractable
Sewage irrigated									
	A_1 0-10	2.4	1.8	3.3	2.6	3.6	2.4	90.00	060
	$A_2$10-20	2.4	2.2	4.8	1.8	3.4	1.9	85.20	0.64
	A_3 20-30	1.9	0.8	2.2	1.2	2.7	1.4	95.00	0.40
	A_4 30-40	1.2	0.3	1.8	0.8	1.1	0.9	90.00	030
Non-sewage irrigated									
	A_1 0-10	1.0	0.75	0.86	0.82	1.01	0.82	76.00	2.00
	A_2 10-20	0.98	0.12	0.28	0.22	0.92	0.30	80.00	0.84
	A_3 20-30	0.22	0.06	0.18	0.18	0.45	0.16	78.00	0.54
	A_4 30-40	0.52	0.02	0.22	0.09	0.38	0.08	76.00	0.72

Benin et al (1981) showed that raw sewage irrigation inhibited plant growth, but did not reduce the yield of wheat grain. The only problem associated with such a practice is the possibility of the accumulation of potentially toxic metals in plants, thereby resulting in their introduction in the food chain. On the other hand, Chakrabarti and Chakrabarti (1988) showed that raw sewage and sludge both reduced the plant growth and yield of grains, but this reduction was more when 66 percent diluted sewage and 20 t/ha sludge were used (Table 101).

Table 101. Effect of sewage irrigation and sludge amendment of soils on vegetable growth and grain yield of wheat crop (Chakrabarti and Chakrabarti, 1988).

Treatment	Vegetative growth (t/ha)	Grain yield (t/ha)
100 % sewage	6.30	2.05
66 % sewage	5.67	1.89
50 % sewage	5.81	1.91
Control (well water)	6.93	2.06
10 t/ha sludge	7.14	2.02
20 t/ha sludge	7.03	2.03
C.D.at 5 %	0.89	NS
C.D.at 1 %	1.24	NS

The use of sewage water was found beneficial for vegetable crops. The application of sewage increases nitrogen and phosphorus levels of soil. Consequently, yield and dry matter production increased with increase of sewage concentration. Water culture experiments (Table 102) show that Cadmium and Zinc contents of plant parts were positively correlated with sewage concentration. This was due to absorption and translocation of these heavy metals. However, in agriculture fields. Cadmium remains on soil. Garcia et al (1976) and Hehne et al (1973) suggested that crystalline inorganic compounds may be formed from heavy metals added to soil or since sewage contains chloride which promotes complex formation, so that uptake of Cadmium was checked. The absorption and translocation of Zinc by *Abelmoschus esculentus* was higher than that of Cadmium.

Amongst six metals (Cd, Cu, Hg, Ni, Pb and Zn), lead nickel and mercury are the least likely to present hazards to animals or man as a result of their consuming forage or food crops grown on land that has been treated with sewage sludge. In general, this is because of the uptake by crops of these three metals is little effected by the addition of sludge. Indeed the potential hazard to lead toxicity in the food chain are reduced whenever it is incorporated into the soil, in contrast to the hazard that may arise from the deposition of atmospheric lead into foliage. Although the concentration of nickel may sometimes be increased in some edible parts of crops, its phytotoxicity appears to provide a safeguard against toxicity to animals and man. The presence of mercury in sewage sludge has caused some concern because of the evidence that inorganic forms of this element undergo microbiological methylation to highly toxic compounds in aquatic systems. However, no methylation has been observed in soils, and Chaney (1973) indicated that hazards from mercury are much less in terrestrial than in aquatic food-chain.

Copper content of crops are not greately increased as a result of sludge application, and it may be noted that the total recovery of added copper in four successive crops averaged only 0.08 percent from two soils that had received 86 kg/ha in a single application of sludge (Kelling et al, 1977). Ruminant animals, especially sheep, are very susceptible to copper toxicity and may be adversely affected by even slight increases in concentration in forage crops caused by application of sludge.

Zinc is normally the most abundant heavy metal in crops, and in sewage sludge of both domestic and industrial origin. Its concentration in crops is increased, sometimes substantially, as a result of addition of sludge. The total recovery of zinc in four successive crops averaged 0.65 percent of an addition of 180 kg/ha in sludge (Kelling et al, 1977). Zinc is more toxic to plants than to animals or human beings. Thus, a concentration of 400 - 500 ug/g in the foliage has been reported as toxic to plants (Chapman, 1966). Whereas, diets containing concentrations of this order have been fed to animals with very little evidence of ill effects (Underwood, 1977).

Table 102. Effect of sewage water on yield and heavy metal contents of *Abelmoschus esculentus* (Shrivastava and Singh, 1990).

Dilution of sewage water v/v	Yield parameter			Heavy metal content (μg/g Dry wt)							
	Dry weight g/plant	Fresh weight of fruits g/plant	No. of seeds/ plant	Cd				Zn			
				Root	Stem	Leaf	Fruit	Root	Stem	Leaf	Fruit
Control	2.8	9.4	14	0	0	0	0	45.0	38.0	65.0	55.0
1:3	3.1	10.7	13	0.6	1.4	1.8	0.4	240.0	295.0	315.0	135.0
1:1	3.3	10.9	16	13	2.0	2.1	0.9	204.0	342.0	325.0	154.0
1:0	5.1	13.6	18	1.4	2.3	2.6	12	165.0	175.0	285.0	178.0

Considerable concern about cadmium as an environmental pollutant has arisen from the suggestion that increased intakes of this metal results in hypertension (Schroeder, 1965), and from the finding that *itai-itai* disease in the Jintsu-Valley in Japan was associated with elevated concentrations of cadmium in some local foods. The concentration of cadmium in crops increased, sometimes by a substantial factor, as a result of addition of sludge. The total recovery of cadmium in four successive crops averaged 0.17 percent of an addition of 4.3 kg/ha in sludge (Kelling et al, 1977).

In view of the uncertainties that remain about the behaviour and effects of cadmium in the food chain, it is desirable to minimise its concentration in crops that are grown on sludge treated land. This may be achieved by using rational management practices, most of which are also applicable to other potentially hazardous heavy metals. It is especially important with cadmium to continuously monitor its concentration in sludge and in crops. Chaney (1973) suggested that the zinc/cadmium ratio in sludge should be greater than 200, and as near as possible to 1000, on the basis that phytotoxicity due to zinc (at 400 - 500 kg/ha) would occur before the cadmium content of the crop constituted a health hazard.

There is no doubt that the whole spectrum of organic and inorganic materials for use of sewage sludge in agriculture must be closely investigated with a view to use them to the best advantage, but at the same time avoid dangers of over manufacturing and metal pollution.

Application of sludge should be made to well drained soil of high electrical conductivity and, perhaps more importantly, the pH of the treated soil should be monitored and adjusted as necessary to atleast 6.5.

EFFECT OF SEWAGE POLLUTION ON WATER BODIES

The discharge of untreated sewage into water bodies can produce some or all of the following symptoms: (a) depletion of oxygen content caused by biological oxidation of organic matter; (b) stimulation of algal growth and also a shift of algal flora to the blue-green algae, leading to the production of obnoxious blooms floating scums or blankets of algae etc. Most of the blooms forming Cyanophytes seem not to be utilised as food by the invertebrates or zooplankton, thus predatory control becomes minimized. Decomposition of such algal masses in turn leads to oxygen depletion (Kumar, 1968).

Planktons have been found to show striking responses to manipulation in their environment. The increase in Phytoplankton density was sudden in early stages and declined in the later stages, Chlorophyll-a concentration corroborated by. concometant increase and decrease in Phytoplankton density (Table 103). Shanon-Weaner's diversity index of Phytoplankton was substantially higher in sewage polluted water in comparison to fresh water without sewage input (Trivedi, 1982 a).

Table 103. Growth of Phytoplankton in sewage treated water (Trivedi, 1982 a).

Date	Parameter	Fresh water	Sewage water
3.12.80	Chlorophyll-a	21.0	21.0
10.12.80	(ug/m^3)	30.0	48-.0
17.12.80		33.0	36.0
24 12.80		40.0	30.0
31.12.80		18.0	15.0
3.12.80	Phytoplankton density	1000	1000
10.12.80	(cells/ml)	1590	1400
17.12.80		1840	1070
24.12.80		1720	860
31.12.80		1603	806
3.12.80	Shanon-Weaner's diversity	0.57	0.57
10.12.80	Index	0.21	1.70
17.12.80		0.42	1.47
24.12.80		0.48	*1.73*
31.12.80		0.52	0.83

In sewage polluted water, the structure of algal communities undergoes several changes. Heterotrophs like *Euglena viridis, Nitzechia palea* and *Oscillatoria limosa* become important. These algae help in supplying the much needed oxygen to the large population of bacteria that breakdown organic matter. In the absence of oxygen anaerobiosis occurs in which fermentation of carbohydrates and sugars and putrification of proteins take plane. In such a breakdown of protein, hydrogen sulphide in produced, which bubbles out giving the smell of rotten eggs. The water body becomes a source of nuisance and fish and other animals are killed. Another kind of synthetic organic matter assuming serious proportion is the detergents. Unlike soap, detergents are not easily decomposed by bacteria and are richer source of phosphate that hasten the process of *eutrophication*. Water quickly gets covered with thick growth of notorious weeds like *Eichhornia, Pistea, Salvinia, Potamogeton, Azolla, Lemna* and *Typha* species (Ambasht, 1981).

HEALTH HAZARDS PROM SEWAGE

Faeces and urine of infected human beings contain all the agents causing infectitious diseases in human beings and animals (Table 104). Infections are spread by the patients and by the carriers who shed the pathogen in faeces or urine. Carriers may be patients suffering further ill health themselves or patients with mild or asymptomatic disease that has neither been discovered nor diagnosed.

There is strong epidemiological evidence that the use of untreated/unstored sewage can be associated with an actual risk of excess infection from a variety of organisms. The smallest are viruses and phages, which can be only visualized under electron microscope (magnification 2000 times). Next ate bacteria, which can be seen under powerful compound microscope (magnification 200 times). Plankton (algae, fungi, protozoa, metazoa) can be easily seen under microscope. Intestinal nematodes, such as, round worm (Ascaris) , whipworm (Trichuris), and hookworm (*Ancyclostoma, Necator*) are visible to the normal human vision.

Table 104. Diseases, causative agents from faeces and urine of infected human beings (WHO, 1972).

Diseases	Causative agents
BACTERIAL DISEASES	
Typhoid and paratyphoid fever	*Salmonella* spp
Illness	*Clostridium perfringens* Type A
Arizona infection	*Arizona arizonae*
Enteropathogenic infection	*Escherichia coli*
Bacillary dysentery	*Shigella sonnel, S. flexneri*
	S. dysenteriae, S. boydii
Cholera	*Vibrio cholerae, V. cholerae*
	biotype "El Tor"
VIRAL DISEASES	
Infectitious hepatitis	Virus A
PARASITIC DISEASES	
Taeniasis	*Taenia sagintata, T. solium,*
Cysticercosis	*T. solium* larva and
	Cystocercus cellulosae
Diphyllobothriasis	*Diphyllobothrium latum*
Pascialopsiasis	*Fascialopsis buski*
Echinostamiasis	*Echinostoma ilocanum*
Clonorchiasis	*Clonorchis sinensis*
Opisthorchiasis	*Opithorchis felineus* and
	O. viverrini
Fascioliasis	*Fasciola hepatica, F. gigantica*
Amoebiasis	*Entamoeba histolytica* cysts
Ascariasis	*Ascaris lumbricoides* eggs
Trichuriasis	*Trichuris trichura*
Enterobiasis	*Enterobius vermicularis*
Balantidiasis	*Balantidium coli*
Gairdiasis	*Giardia lamblia*
Isospora infection	*Isospora hominis, I. belli*
Dietamoeba infection	*Dietamoeba fragilis*

The actual risk of excess of infection by bacterial and viral diseases appear to be lower than for nematodes infections. The epidemiological evidence for this is relatively weak, owing to a paucity of credible epidemiological investigations, but the assumption is based on sound theoretical considerations of potential risk (Strauss, 1990; Shuval, 1986; Blum and Feachem, 1985).

VIRUSES

Certain viruses that multiply in the alimentary tract of human being and may be excreted in considerable amounts in faeces can be found in sewage and polluted water. The viruses most commonly present in sewage are the Enteroviruses (*Poliovirus*, *Coxsackieviruses* and *Echoviruses*), Acenoviruses, Reoviruses and the virus of the infectitious hepatitis (Chang, 1968). A striking example was the epidemic of infectitious hepatitis in Delhi (1955-1956) in which more than 28000 cases were identified, with a case fatality rate of 0.9 per 1000 and an estimated total case incidence of 97000 (Viswanathan, 1957).

BACTERIA

They are of 0.5 to 1.0 u size and of different shape such as cocci, bacilli, spirillum type. They are either saprophytes or parasites. The former type live on the decaying organic matter which is abundant in sewage, while the later are accustomed to organic matter found in the body of organisms and do not live long in sewage. Like all organisms the bacteria are quite sensitive to their environment, especially temperature and pH. Normally, the bacteria grow faster at higher temperature and hence we find that the bacterial contents of sewage is high in summer and low in winter.

Pathogenic bacteria are transmitted directly by water or indirectly through water to food, and constitute one of the principal sources of mortality and morbidity in many developing countries. They include the causative agents of the great epidemic diseases - Cholera and Typhoid - and of the less spectacular but numerous cases of Infentile diarrhoea, Dysentry and other Enteric infections that occur continuously and often with fatal results among the rural and urban populations. Observations (Table 105)

Table 105. Diseases of human beings caused from the faeces of animals (WHO, 1972).

Diseases	Causatiye agent	Source
Salmonelloses	Salmonella app	Domestic & Wild
Botulism	*Clostridium botulinum* and C. *parabotulinum*	Animals
Arizona infection	Arizona arizonae	Animals
Bolivian	Machupa virus	Rodent urine
Q fever	*Coxiella burnetti*	Cattle, sheep, Goat
Diphyllobothriasis	*Diphyllobothrium latum*	Dogs
Sparganosis	Sparganum of *D. latum and Spironensis*	Cat and Dog faeces
Angiostrongyliasis	*Angiostrongylus cantonsis*	Rat faeces
Anisakiasis	*Anisakia spp*	Fish eating mammals
Fasciolopsiasis	*Fasciolopsis buski*	Dog/Hog
Echinostomasis	*Echinostoma ilocanum*	Dog, Rat
Clonorchiasis	*Clonorchis sinensis*	Cat, Dog, Hog
Opisthorchiasis	*Opisthorchis felineus*	Fish eating mammals
Fascioliasis	*Fasciola hepatica F. gigantica*	Sheep, Cattle, Herbivores, Omnivores.
Heterophyid infection	*Heterophyes heterophyes*	Fish eating birds and mammals.
Metagonimiasis	*Paragonimus westermani*	Carnovores animals
Hymenolepasis	*Hymenolepis diminuta*	Rat, Mice
Gnathostomiasis	*Gnathostoma spinigerum*	Dogs and cats
Echicoccosis	*Echinococcus granulosus*	Dogs and wolf
Toxoplasmosis	*Toxoplasma gondii*	Cat

shows major diseases attributable to the ingestion of water borne bacteria. During the last three decades, classic Cholera caused by *Vibrio cholerae* has recorded remarkably at Calcutta (WHO, 1972). Typhoid and Hepatitis assumed an alarming form in 1980 in Maharashtra state where sewage water got mixed up with drinking water at Sherinala in Sangli, Chinchward in Pune and in the southern district of Bombay (Deoras, 1981).

ALGAE

Unicellular algae belonging to Cyanophyceae and Chlorophyceae present in the sewage help them to utilize radiant energy and convert mineral food matter (nitrogen sulphur etc.) into organic tissues.

FUNGI

They grow luxuriently in sewage where they find ample food. Since they do not need any radient energy from sun, they can grow deeper depth and in turbid water.

PROTOZOA

They are small unicellular animals of irregular shape and size such as Amoeboids (*Saricodina*), Flagellates (Mastigophora) and Ciliates (Infusorial). The protozoa may be again holozoic, saprozoic or holophytic. The holozoic protozoa utilize solid food from plants and animals. The saprozoic protozoa utilize carbohydrates and proteins that have been made soluble by bacteria and other organisms. The holozoic protozoa contains chlorophyll or other pigments and absorb radiant energy to convert minerals into organic tissues.

Among the protozona *Entemoaeba histolytica* causes intestinal amoebiases and amoebic liver abscess. It is wide spread throughout the tropical countries and wherever sanitory conditions are poor. Amoebic cysts are resistant to chlorine in dozes normally applied in water (WHO, 1969).

Besides the diseases of human beings caused by pathogenic agents in the faces and urine of human beings, a large number of other diseases are caused by pathogenic organisms from the faeces of animals (Table 105).

The Guinea worm which causes Dracontiasis, is common among the rural population. Some intestinal Helminths such as *Ascaris lumbricoides* and *Trichuris trichiura* may be water borne. Hydatid disease (Hydatidosis) a zoonosis that usually involves a dog-sheep-dog cycle in maintaining the reservoir of infection, is occasionally transmitted to human beings through drinking water or food contaminated with excreta of the primary hosts (WHO, 1972). In polluted water reservoirs and streams various parasites and vectors of parasitic zoonoses (*Diphyllobothrium, Fasciolopsis, Schistosoma, Heterophyes, Dracunculus* etc.) may develop and persist. The parasito-logical problems associated with sewage have become important in almost every part of the world. Soil pollution, primarily by eggs and larvae of different Helminths and by Protozoan cysts, is a scourage of some rural and urban areas. This a particularly hazardous where night soil or inadequately treated sewage sludge is used for manuring crops (WHO, 1980).

SEWAGE DISPOSAL

In many towns of India, there are no facilities for efficient discharge of sewage (Raj, 1977). It is usually carried through open drains/ masonary canals to meet the neighbouring water bodies, or directly disposed of near the water bodies, or in the water bodies. Table 106 gives an idea about the magnitude of this problem even in our big cities.

Animal faeces which often contain pathogens pollute pavements, streets, lawns, public paths, play, grounds, sand pits and the floor of houses. In our traditional system of waste disposal in heaps of solid manure away from houses, the spontaneous generation of heat kills pathogenic organisms. However, they survive in water suspension called "slurry" in intensive units (WHO, 1978).

Our fresh water bodies are also polluted by the cremation methods in practice. It has been estimated that 25000 to 30000 dead bodies are cremated at Varanasi burning ghats annually. Out of the total dead bodies 40 percent come from out of Varanasi district due to the religious importance and 60 percent from Varanasi town. Floating human dead bodies and animal carcasses add another dimension of pollution.

Table 106. Status of sewage disposal in some of the major cities in India (Raj, 1977).

City	Population	Present practice of disposal
Calcutta	75,00,000	Only a part of the city is sewered and primary treatment provided. Part used for farming. Balance to river in untraeted condition.
Bombay	55,00,000	Only 1 percent of the city sewage is treated partly with secondary treatment. Balance goes untreated to the sea, and an odour problem exist io some areas.
Delhi	36,00,000	Three modern plants with primary treatment for 98 mgd and secondary treatment for 32 mgd.
Madras	25,00,000	Part of sewage used for irrigation and part goes direct to the sea.
Kanpur	15,00,000	No sewage treatment plant. Part of 6 mgd used for irrigation.
Hyderabed	13,00,000	Screening and grit removal followed by irrigation.
Bangalore	12,50,000	No sewage treatment plant.
Ahmedabad	12,50,000	Two modern sewage treatment plant.
Poona	7,50,000	
Nagpur	7,00,000	
Lucknow	6,75,000	
Agra	5,50,000	No treatment. Pumped for irrigation.
Varanasi	5,00,000	
Madurai	4,50,000	
Allahabad	4,50,000	
Amritsar	4,00,000	
Indore	4,00,000	Secondary treatment plants.
Jaipur	4,00,000	
Sholapur	3,50,000	
Kota	3,00,000	No treatment plant. Goes to river Chambal
Kolhapur	3,40,000	No treatment. Pumped for drinking.

Many of our lakes are becoming increasingly murky, smelly and chocked with excessive growth of algae. Most of the rivers become darkened with sewage and other undesirable foreign and extraneous matter. The river Damodar is perhaps the most heavily abused as a convenient dumping ground for wastes and effluents of all kinds, including agricultural, industrial and domestic. Many parts of the river do not seem to have any dissolved oxygen and no wonder they fail to support the growth of any desirable aquatic organisms. Only such organisms as Bloodworms and certain other undesirable animals, pathogenic bacteria and pests are found (Kumar, 1977).

FLUSHLESS, ODOURLESS TOILET

A flushless toilet launched in Britain requires no disinfectants, no connection to sewers, and turns its wastes into compost. To prevent it smelling, the toilet relies simply on an electric fan. Peter Soulsby of Eco-Clear, the company selling the lavatory, estimates that if it replaced every toilet in England and Wales, it would save 3 billion litres of water a day, a quarter of all daily consumption.

The pan resembles a normal one, but it is made from stainless steel and sits atop a chute 35 centimeters across. The general idea is that not too much touches the sides, the pan would require periodic cleaning. Debris passes down the chute and lands on a bed of sawdust contained in a large polyethylene tank sited underneath (Figure 32). Here the debris slowly decomposes into compost, The first solid compost would probably take around 2 years to appear.

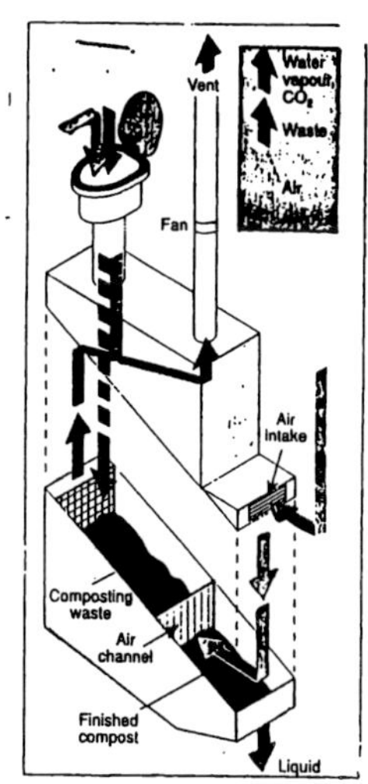

Figure 32. Flushless, odourless toilet for water starved location.

The largest tanks the company produces cope with 45,000 uses of the toilet per year, but even these would produce only a few bucketfulls of solid compost and 4000 litres of liquid compost annually.

Much less water is used in this system, the Clivus toilet. Some 360,000 litres of water disappear down the conventional toilet if it is used 40,000 times, compared with just 2500 litres for a clivus system.

The unpleasant odours normally produced by lavatories and sewage originate from the action of anaerobic bacteria on waste. These bacteria, which survive on airless conditions, generate foul-smelling products such as mercaptans and other sulphurous compounds, and methane gas which is flammable.

The Clivus toilet overcomes this problem with an electric fan which draws air through the compost pile. This eliminates odours, because only bacteria that breathe air can survive in the compost pile. These aerobic bacteria generate odourless carbon dioxide and water vapour as they degrade the waste, and these are flushed into the atmosphere by the fan.

SEWAGE TREATMENT

The foregoing description clearly reveals that sewage contains high amount of pollutants and requires suitable recycling/treatment prior to its discharge in the water bodies or on the soil to restore the ecological balance in the aquatic or terrestrial ecosystems.

In recent years in India, the construction of sewage treatment system have been pushed very strongly by the government. Sewage is being considered as one of the largest field of public finance. Underground sewage works/ a motor car, and Jet Plane has become the 'Status symbol' in the last few decades in our country. In some urban towns underground system for carrying sewage have been layed down in the past, however, only few of them have been able to maintain the treatment plants satisfactorily. Such huge projects are meant for only a part of the community, and have not been able to serve the whole population (Mehta, 1978). This is due to the fact, that present dominant sewage technology with huge scale sewage works has a typical inappropriate character of modern technology - capital intensive, energy wasting, expert dependent, ineffective and bringing profits to a small proportion of the society and leaving environmental and economic burden on the rest of the society. Ui (1980) has rightly said that 'construction of this type of sewage treatment units should be stopped immediately and the whole plan should be reviewed critically.'

Experience from other developing countries in Asia, like Japan have however, shown that from the expenditure point of view, smaller scale of treatment plants and smaller pipes underground in shallow will make finally cheaper cost, even though there will be comparatively numerous works and pipes.

Our large scale sewage treatment plant, once constructed, has a weak point that it can never carry on its work, when out of order. But the numerous small scale treatment works, it never happened like the case of big once. Because even when one is out of order the rest are still working as usual. The big sewage treatment plants have a tendency towards unlimited increase of components such as parts and fittings. These are the main causes of the trouble. Once a trouble started incidently, the big plant is to loose its functioning for 50 to 60 days atleast. Hence instead of going for huge sewage treatment plants, we should plan to have small scale sewage treatment plants, to save the capital cost of installation and maintenance.

We have not correctly realized that we are already short of electricity (Mehta, 1968). It is recommended that we should take the advantage of wind power to stir-up the waste of the oxidation pond. Wind mills are being used at Isunigava Factory of Takizawa Ham Company, Japan for extended aeration. Ammonia contained in the sewage can be changed biologically by nitric acid, respond to the changing of wind power energy. This process will be takenup as reverse of chemical energy. When the wind blowing is ceased, the chemically changed nitric acid, instead of air, can be a help for the breaking of bacteria which putrify the water in the pond. This theory and practice, if accompanied, makes the sewage clean by the power of nature itself (Ui, 1980).

If we measure oxygen level in the mixed liquor and cut off energy supply for agitation of it according to the increase of wind energy, we can reduce the electrical energy input for the sewage treatment. This is possible with the application of oxygen electrode in the mixed liquor.

SEWAGE RECYCLING

Recycling of excreta and waste water has been practiced for many centuries. Particularly in Asia, excreta are important in agriculture, fish farming and aquatic vegetable production. In rural China, for example, excreta are used as soil conditioner and fertilizer. In the hill areas of Java (Indonesia), excreta are traditionally used to fertilise fish ponds. Some of the excreta collected from household bucket laterines in Indian cities are applied by farmers to nearby fields (Strauss, 1990).

The use of sewage in agriculture and fish farming began with the development of flush laterines and sewerage technology more than a century ago. Waste water from many European cities was applied to nearby farmlands, this being the most economic method of disposal. However as the amount of wastewater increased, availability of land for this purpose decreased. The agricultural use of wastewater is gaining popularity, particularly in arid and semiarid regions, for the following reasons:

- Scarcity of water for irrigation;
- Increased need for domestic water;
- Expansion of agricultural production;
- High cost of mineral fertilizers;
- Few health risks and little damage to soil if precautions are taken.

Sewage is used notably in India, Latin America, Tunisia, United States of America and West Asia. Practices vary from use of raw sewage to well treated sewage (Table 107).

In China, the custom of storing human waste from rural and urban populations and using it in agriculture is still practiced despite industrialization, urbanisation and the use of chemical fertilisers. Excreta collected in urban areas are transported to rural areas by trucks, tractors, cart and boat. In 1981, out of the 73 million tonnes, 40 million tonnes were used in agriculture, fish farming and aquatic vegetable production. It has been estimated that of the 150 million tonnes of faecal waste produced annually by some 800 million people in rural China, 60-70 percent is recycled. In some places excreta are stored for several weeks before use in order to destroy eggs. Co-composting of human,

Table 107. Some ways in which wastewater and excreta are used

Continent/ country	Town or area	Waste used	Crops irrigated/produced	Approximate area irrigated (ha)
1	2	3	4	5
The Americas				
- Argentina	Mendoza (city)	Primary effluent (at times diluted) and dried sludge from sewage works	Lettuce, onions, tomatoes, artichokes	2000
- Chile	Santiago (Zanjon de la Aguada)	70% of city's wastewater (untreated or at times diluted by storm water runoff)	Lettuce, cabbage, celery, cereals, grapes	6000
- Poru	Country as a whole	Wastewater (treated, partly treated or untreated)	Miscellaneous edible and non edible crops	5000-6000
	San Martin de Porras (Lima)	Wastewater (untreated)	Tomatoes, radish, spinach fodder and non-edible crops	1500-2000
	lca (Cachiche)	Primary pond effluent	Cotton, maize, grapes	400

1	2	3	4	5
- Guatemala	Rural areas	Stored excreta from dry, double-vault latrines	Maize, miscellaneous vegetables	Not known
- Mexico	Mezquital Valley (State of Hidalgo), Rural Development District 063	Wastewater from Mexico City (untreated, diluted or impounded)	Maize, wheat, oats, green tomatoes, chillies, fodder	58000
	Mexico City	Secondary effluent	Parks and green belts	Not known
USA (California)	Bakersfield	Wastewater: effluent from aerated lagoons	Barley, field corn, cotton, pasture	2000
	Irvine	Wastewater: effluent from tertiary treatment plant (includes filtration and chlorination)	Tomatoes, chillies, asparagus, broccoli, cauliflower, sweet corn, citrus orchards	> 5000

1	2	3	4	5
Europe				
- Federal Republic of Germany	Braunschweig	Wastewater (mixed 90% secondary effluent +10% raw) & sewage sludge	Cereals, sugar beets potatoes	2800
-United Kingdom	Greater London	Lagooned, digested sludge from sewage works	Field and horticultural crops	($3x10^6$t of sludge annually at 6-10% solids)
Northern Africa and Western Asia				
- Tunisia	Tunis Soukra	Wastewater (secondary effluent)	Citrus tres	600
- Jordan	Zarga river/ King Talal Dam basin	River water consisting mainly of treated sewage	Trees, industrial crops, vegetables eaten cooked, vegetables eaten raw, depending on crop restriction	500

1	2	3	4	5
- Kuwait	Country as a whole	Wastewater (tertiary treatment plant effluent)	Fruit trees, fodder crops, maize, wheat, vegetables (eaten raw or cooked); forestry	1000-2000 (estimated on basis of installed irrigation pump capacity)
- Saudi Arabia	Riyadh-Dirab	Wastewater (tertiary treatment plant effluent)	Wheat, fodder crops	2500
	Riyadh Dariyah	Ditto	Date palms, lemon trees, fodder crops	800
South-East Asia				
- China	Country as a whole	Excreta (untreated or stored or co-composted with animal manure and crop residues)	Reportedly, up to 70% of the excreta pro- urban night-soil are used as basic soil Application is usually by ploughing or apparently not restricted to certain crop communally.	

1	2	3	4	5
- India	Country as a whole	Wastewater (untreated, diluted; partially treated)	Paddy rice, maize, wheat, sorghum, vegetables, fodder crops	> 70 000
	Calcutta	Wastewater ponds	Fish (carp species, tilapia, catfish)	3000
- Indonesia	Java Highlands	Excreta-fed domestic ponds	Fish (carp and tilapia) and water vegetables	21000

and animal excreta with crop residues is also practiced. About 1.8 billion tonnes of organic fertiliser are produced annually by these processes. Its nutrient value is estimated to be equivalent to that of 100 million tonnes of organic fertiliser.

Most human and animal excreta, and excreta-derived compost is generally ploughed or harrowed into the soil before crops are sown. The application rate for compost is usually in the range of 100-300 tonnes/ha/year; that for liquid night soil is generally 20-30 tonnes/ha/year.

Excreta and waste water must be either stored long enough or treated adequately before use. The longest surviving pathogen are worm eggs, particularly those of the round worms *Ascaris,* which may remain viable for many months or even a few years. *Salmonella* may survive for few months. Time and temperature are the two most important factors determining the storage periods for excreta at ambient temperature in warm climates and indicates the various degrees of hygenic quality involved.

Table 108. Recommended storage period for excreta at ambient temperature in warm climates (Strauss, 1990).

Storage period	Hygienic quality achieved
> 2 days	Inactivation of *Clonorchis* and *Opisthorchis* (liver fluke) eggs·
> 1 month	Inactivation of viruses, protozoa, bacteria (except possibly *Salmonella* on moist, shaded soil) and schistosome eggs
>4 months	Inactivation of roundworm eggs, e.g., hookworm and whipworm (*Trichuris*); survival of some Ascaris eggs
>12 months	Complete inactivation of *Ascaris* eggs

Storage periods of several months to a few years are required at ambient temperatures, but treatment periods can be reduced proportionally to achieve hygienically safe products at higher temperatures. With thermophilic composting, for example at temperatures of 60-70°C, safety can be achieved within a few weeks.

SEWAGE STABILIZATION

The environmental impacts of sewage farms can be alleviated by following sound farm management techniques and using treated sewage for irrigation. In tropical countries, the retention of sewage in stabilization ponds is often the most feasible method of removing pathogens. The wastewater flows through several ponds in three to four weeks. Helminths eggs and Protozoan cysts are removed by sedimentation, while bacteria are virus die (Strauss, 1990).

The treatment of sewage in stabilization ponds has been widely adopted in USA, Canada, Australia and New Zealand. In India, one experimental pond was designed by Raman (1960) at Madras. The pond was open earth digged one, about half an acre and about two feet in depth, for a population of 800, amounting on the average to 40000 gallons per day, was pumped into one corner of the pond, where a small area was bounded off to form a priliminary settling zone for the sewage with a detention capacity of about 1.8 days. The settled sewage from this preliminary pond was allowed to flow through submerged ports in the dividing bund into a secondary pond with a surface area of 20000 sq.ft., and a detention capacity of nine days. The later zone served as a stabilization pond for the study (Figure 33).

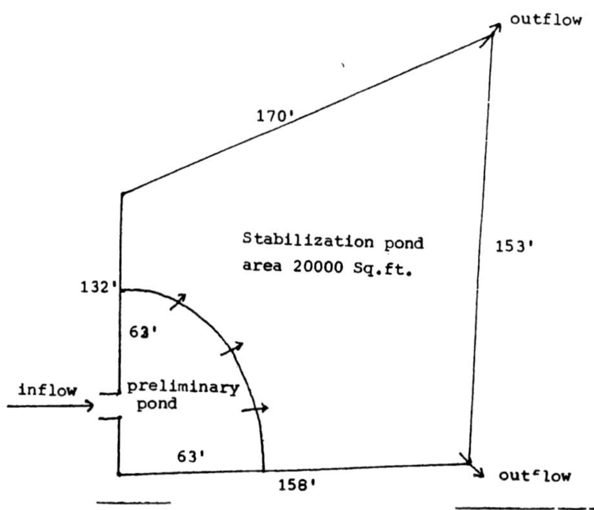

Figure 33. Sewage treatment pond (Raman, 1960).

Stabilization pond are easy to construct, and are self-regulating, and require practically no supervision, when sufficient land is available. They form a cheap and efficient method of sewage treatment. This method, seems to be well suited for our conditions because of the favourable factors like abundant sunlight and warm climates. From an aesthetic point of view, the stabilization pond did not have any disagreeable features. Sewage odours were never experienced. Floating scum of sewage solids, emergent vegetation and mosquitoes breeding were no problems.

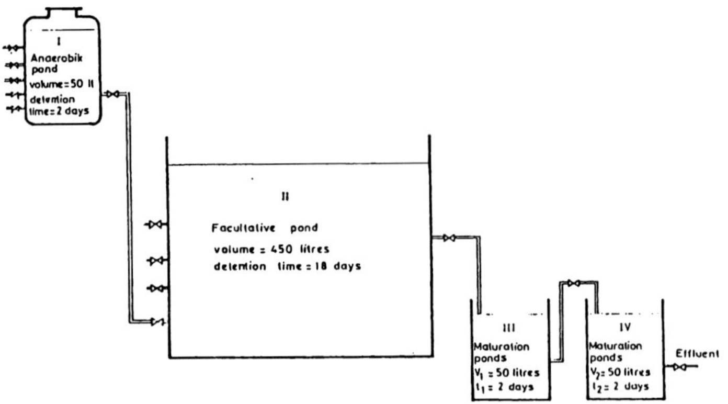

Figure 34. Pilot stabilization ponds model (Sengul and Turkman, 1988).

A pilot mode was prepared in order to simulate the detention time of Izmir Municipal wastewater stabilization treatment plant which was proposed for the city of Izmir, Turkey.

The pilot treatment plant consisted of three main units in the order of anaerobic, facultative and maturation ponds and they were arranged to work as semi-centinuous reactors. The volumes of anaerobic, facultative and maturation ponds were 50, 450 and 100 litres, and the detention times of these units were 2, 18 and 4 days respectively. By arranging the volume of feed water, the detention times for the three units mentioned above-was established. Pilot stabilization pond was under operation between 1985 to November 1987 (Figure 34).

The pilot treatment plant was fed daily with domestic wastewater taken from Bornova Sewarage system. 25 litres of sewage was fed into the model and 25 litres of treated water was discharged every day.

SMALL SEWAGE TREATMENT UNIT

Ui (1977) developed a small unit of sewage and other organic waste water treatment, from the experience of operation of oxidation ditches, which has been used popularly in Europe, especially in Holland. The unit consists of one or more shallow ponds of 2 meter depth, agitated intermittently by turbine on suitable floats, to mix the contents of the ponds, as well as supply of air to the activated sludge in the ponds. There are several plants of this type in operation in Japan for the treatment of hog waste and food processing factory waste. Similar to oxidation ditch, intermittent operation of turbines makes the same pond to function as aeration and sedimentation. In case of nitrogen containing waste water, the pond has both function of nitrification and denitrification. With such small sewage treatment unit, more than 90 percent removal of organic materials in terms of BOD, and more than 70 percent removal of ammonia and nitric nitrogen has been achieved for hog waste, which has 10 times stronger loading than domestic sewage. There is no need of sophisticated technology for the building of small sewage treatment units. The agitating turbines are made from the parts of used automobiles, its operation also do not need specific skills. This process is

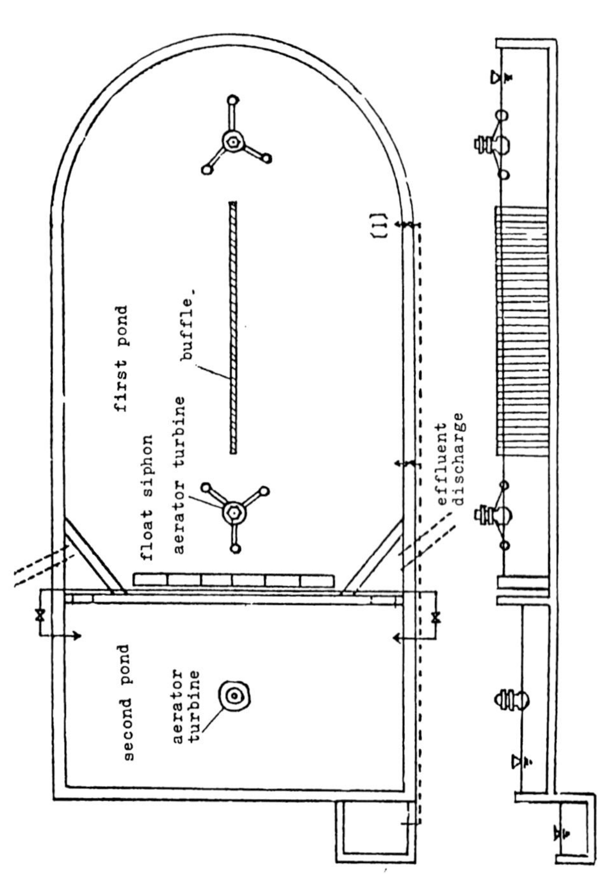

Figure 35. Small sewage treatment units.

suitable for 100-10000 persons unit, or 10-1000 pigs units, and has a merit of odourless operation. The piping cost for the collection of sewage for such small unit is also small, so the total cost for the sewage system is much smaller than the large scale units which are under construction everywhere now.

When such small units of sewage treatment is planned, built and operated by the local people, it is in fact a part of local development plan by the people. On the contrary, the construction of huge sewage treatment units with enormous quantity of steel and cement are handled by experts, and involve huge capital investment and demand for electricity. It is quite clear, which way is more appropriate for human being.

MODERN SEWAGE TREATMENT

Public health engineers are aware of the working of some of the modern Tertiary Tretament Plants producing purified waste water including the much publicised South Tahae Water Reclamation Plant in the United States of America and the Windhoek Water Reclamation Plant in South Africa. Aother frequently quoted example is of Bethlehem Steel at Sparrow Point, Maryland, USA, where the sewage from Baltimore city is purchased for reclamation and reuse. Starting with 25 million gallons per day (mgd) in 1942, the Company had increased its demand to more than 125 mgd in 1966. Dow Chemicals Co. is also using about 6 mgd of purified wastewater for its factory at Midland, Michigan.

In Bombay, raw sewage after Tertiary treatment has been used for industrial purposes. This has been done at Chembur, a Bombay suberb, by Union Carbide for use in their plant and in many multistoryed buildings for air-conditioning. The modern tertiary sewage treatment plant of 1 mgd (4.540 million litres per day, mld). Besides the normal sewage treatment which generally forms part of complete treatment of sewage for discharge into drains or small rivers that is upto the final clarifier (Unit 6). The treatment has proceeded further is a clarifloculator, a rapid sound failter, softener, and then followed by post-chloronation (Figure 36).

The later steps are generally what are adopted in a modern water treatment plant and thus the Union Carbide Plant is a combination of a secondary sewage treatment plant plus a water plant. The price of treated and softened water is about 65 paisa per 1000 litres (1972) as compared to Re 1.00 per 1000 litres of Municipal water.

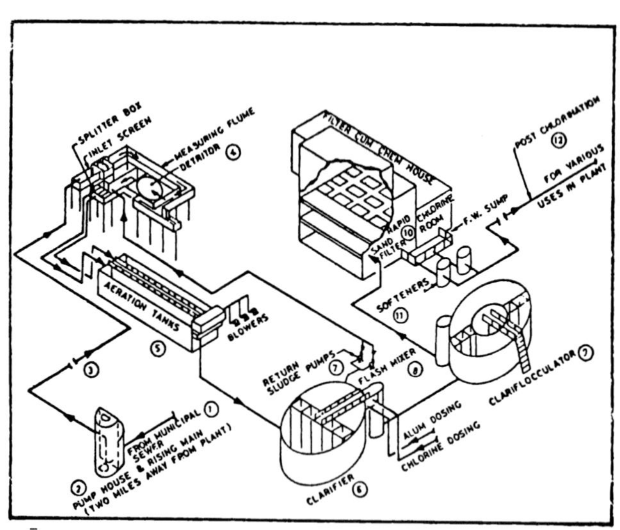

Fig. 36. 4.54 MLD tertiary treatment plant at Union Carbide Co. Ltd., Chembur (Bombay) (Mehta 1975).

(1) Source of sewage, municipal sewer.

(2) Pump House.

(3) Rising Main

(4) Measuring flume and detritor.

(5) Aeration tank.

(6) Clarifier.

(7) Return sludge pumps.

(8) Chemical dosing equipment for addition of alum and pre-chlorination.

(9) Clarifocculator.

(10) Rapid sand filters with back-wash equipment.

(11) Softening.

(12) Post-Chlorination.

The effluent is softened because of process requirement and about 15 percent of total water, which is used for boilers, is demineralized at an extra cost. The analysis of raw and treated effluent (Table 109) indicate that there is little difference between the Bombay Municipal water characteristics and reclaimed sewage, except that the dissolved solids are high.

Table 109. Quality of raw sewage, treated water and water from Municipal main.

Particulars	Raw sewage	Treated water softened & chlorinated	Water from Municipal main
pH	7.15-7.65	7.1	7.8
Total hardness	120-162 mg/L	140 mg/L	39 mg/L
Alkalinity	125-200 "	135 "	40 "
Chlorides	96-130 "	60 "	18 "
Total dissolved solids	550-640 "	300 "	90 "
Suspended solids	70-100 "	NIL	10 "
Turbidity	-	3 "	6 "
BOD at 20°G	220-250 "	1 "	1.2 "
COD	250-350 "	3.5 "	1.7 "

Source: Union Carbide Co Ltd, Chembur

It should be emphasized that all sewage treatment process aim at removal or organic matter, both dissolved and suspended, and inorganic only in the suspended form. The dissolved solids cannot be removed except by very costly methods and thus the reuse can only be done once or twice. If dissolved mineral salts are to be removed, softening is needed (for removal of hardness) and demineralisation for removal of inorganic dissolved salts. Cases are on record, where in order to use the sewage effluents for domestic water supply or rather for charging the water aquifers the secondary sewage effluents have been injected into the ground for recharging the water aquifer. Isreal has done considerable work, and so have some of the dry areas of the USA such as California and Arizona. It should be emphasized that to prevent buildup of dissolved salts beyond limit of potable water, only a part of the sewage can be recycled.

NON CONVENTIONAL SEWAGE TREATMENT PLANT

A non-conventional sewage treatment plant, much cheaper and compact than the conventional one is quite feasible. Raw sewage is led to a settling process where solids are separated from liquid. The solids go to the digester to produce methane gas and digested solids. The methane gas is fed to gas turbine to generate electrical power for running the factory as well as to sell the surplus to other consumers. The digested solid is sold as manure. The liquid from the settling process goes to biological aerator where more solids are produced and separated to be fed to digester. The liquid or water from the biological aerator is clean enough to be used for aquaculture where algae and fishes are grown. Algae is harvested as poultry feed and fishes are marketed. The water from aquaculture flows out as irrigation water. Thus, each sewage treatment plant functions as a resource recycling unit, producing energy, manure, poultry feed, fish and irrigant. Hence, efficient marketing of these products is also conceived as an integral part of resource recycling unit. A conceptual model based on the treatment plant developed by the National Botanical Research Institute, Lucknow is presented in Figure 37·

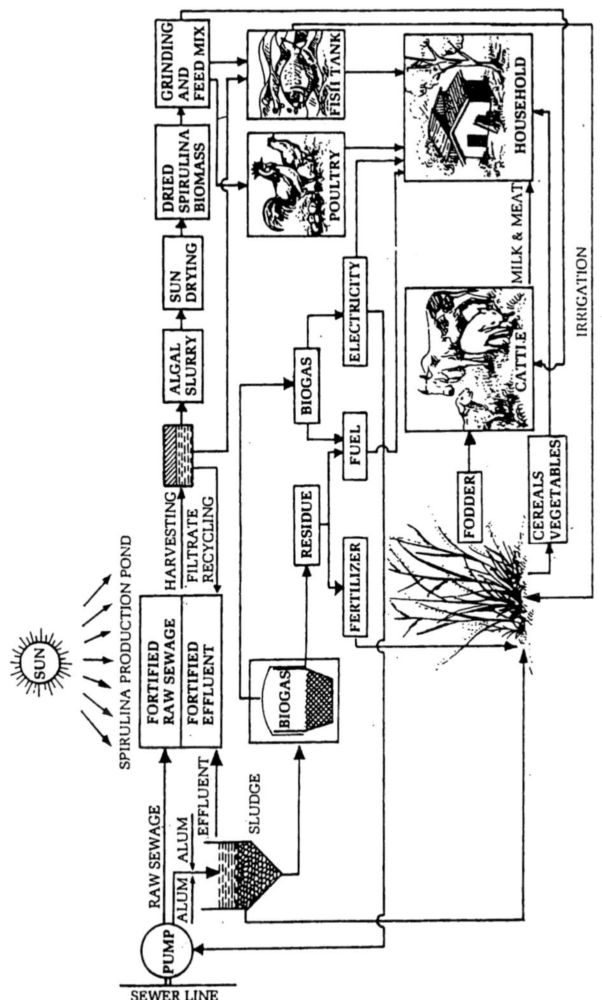

Figure 37 . NBRI-NPJN integrated sewage utilization system.

Chapter-8

AQUATIC PATHOGENS AND NUISANCE ORGANISMS

The natural waters harbour a diverse aquatic life which can range from micro-organisms such as bacteria and algae to higher organisms including vertibrates like fishes. In polluted waters, the density of these organisms usually remain low because of the limited availability of food and nutrients. On the other hand, polluted waters, especially, the eutrophie ones, have a variety of micro-organisms and other biota. Eutrophic waters are rich in nutrients, which promote the algal growth and allow a large number of predators to feed upon it. The presence of organic matter in substantial quantity in these waters also increases the abundance of saprophytic bacteria.

Polluted waters usually harbour a characteristic population of flora and fauna. The sewage contaminated water may also contain various pathogenic organisms belonging to bacteria, viruses, protozoa etc. in addition to the usually innocuous populations. The consumption of these waters, containing the faecal matter, may result in severe health hazards. Most human intestinal pathogens do not survive for extended periods outside the body of the host, but they can remain sufficiently viable in aquatic environments to infect human beings. The infected persons usually excrete large number of these pathogens in the urine and faeces, which ultimately get their way into the municipal sewage. The sewage can contaminate the water resources from where the public water supply is drawn. The contamination of water can also occur in the distribution system through leaks. The human health may be affected by ingesting water directly or in food, by using it in personal hygiene or for agriculture, industry or recreation, and by living near it. Two main categories of water associated health hazards have been considered here: (1) hazards from biological agents that may affect human beings followed by ingestion of water, (2) other forms of water contact other than ingestion, and (3) hazards of diseases transmitted by water associated insect vectors.

1. WATER ASSOCIATED HAZARDS FROM INGESTION OF BIOLOGICAL AGENTS

The principal biological agents transmitted in this way can be grouped into the following categories: pathogenic bacteria, viruses and parasites can be attributed either to the pollution of the water source itself or to the pollution of the water during its conveyance from source to consumer. The pollutants may include the excretions, faecal and urinary, of man and animals, sewage and sewage effluents, and washings from the soil. Infections are spread both by patients and by carriers who shed the pathogen in faeces or urine. Carriers may be patients who have recovered but still harbour the infective agent without suffering any further ill-health themselves, or patients with mild or asymptomatic disease that has neither been discovered nor diagnosed.

(a) Pathogenic bacteria

Pathogenic bacteria transmitted directly by water or indirectly through water to food, constitutes one of the principal sources of morbidity and mortality in many developing countries (Table 110). They include the causative agents of the great epidemic diseases - cholera and typhoid - and the less spectacular but far more numerous cases of infantile diarrhoea, dysenteries and other enteric infections that occur continuously, and often with fatal results, among rural and urban populations, particularly in the developing countries.

Table 110. Human diseases through ingestion of animate pathogens in contaminated water.

Diseases	Causative organism
Cholera	*Vibrio Cholera*
Bacillary dysentry	*Shigella spp*
Typhoid	*Salmonella typhi*
Paratyphoid	*Salmonella paratyphi* A, B, C.
Gastroenteritis	Other Salmonella types, *Shigella. Proteus* spp.
Infentile diorrhia	*Enetropathogenic* types of *Escher ichia coli*
Leptospirosis	*Leptospira* spp
Tularaemia	*Pasteurella tularensis*
Infectitius hepatitis	Virus not yet identified
Intestinal amoebiasis	*Entamoeba histolytica*
Dracontiasis	*Guinea worm*
Distomatosis	*Fasciola* spp., *Dicrocoelium* spp.
Ascariosis	*Ascaris lumbricoides.*

The bacterial species of the genus *Salmonella* are one of the common species implicated in waterborne diseases of human beings. Some species also cause infections in animals. The pathogens are transmitted directly by consumption of contaminated water or shellfish grown in contaminated waters.These bacteria can survive for considerable distances in the downstream after their discharge in rivers. Some of the common infections of *Salmonella* and typhoid (*S. typhi*), paratyphoid (*S. paratyphi*), gastroenterites and food poisoning. As the *Salmonella* is a common cause of food poisoning, it causes global problems because of the international trade of large quantities of food products, animal and animal feed. Substantial proportion of this material may be contaminated. This has necessitated a surveillance of *Salmonella* during international trade and travel, which has been coordinated by the world health organisation.

Vibrio cholerae causes Cholera which is an acute epidemic intestinal disease. The bacteria produces an exotoxin which causes the gut cells to produce excess water along with Sodium carbonate and Potassium. The disease is characterised by the production of severe diarrhoea eliminating "rice water" stools in large quantities, which may cause death within few hours due to severe dehydration. The occurrence of Cholera caused by *V. cholerae* has considerably declined, but its biotype "El Tor", which also produces the same symptoms, is on spread causing several out breaks of Cholera in many countries. The epidemic of Cholera are quite common during the floods and disasters like earthquakes, famines and wars.

Species of *Shigella* (*S. sonnei, S. flexneri, S. shigae*) cause bacillary dysentery. The survival of these bacteria is low in water, but the disease can also be transmitted by housefly carrying infection from faeces to the food, infentile diarrhoea and gastroenterites in infants is caused by entero-pathogenic bacteria *Escherichia coli*. Some urinary infections are also caused by the strains of *E. coli*, but the role of water is that is of little significance as most of them have been found to be caused by individuals own intestinal flora. *Leptospira* spp are this and spiral in shape, which are responsible for causing severe ailments in human beings, infecting liver, kidneys, and central nervous system. The infection is commonly called Leptospirosis. The primary hosts for numerous pathogenic species of this genera are mainly rats, dogs, foxes, volves, and cats from where the contamination reaches to water. Human beings may become infected by drinking or by swimming in contaminated water. *Leptospira canicola* causes Canicola fever which is characterised by high fever, headache, severe muscular pain, conjunctival congestion, jaundice and vomiting. *Laptospira icterohaemorrhagiae* causes Weil's disease, a typical jaundice with fever. The disease is quite common in sewage workers.

In India all the major river waters have become polluted with faecal coliform bacteria (Table 11). These bacteria were observed to be maximum in the Mahi river followed by Narmada river and Tapti rive. These values were minimum for Godavari river.

Table 111. Faecal coliform contamination of river water.

River	Number/100 ml)
Mahi	550,000
Narmada	260,000
Tapti	37,000
Wainganga	3,699
Cauvery	439
Krishna	57
Godavari	7
Periyar	767
Sabarmati	1,174

Pasteurella tularensis causes a disease called Tulareamia, which is characterised by severe constitutional upset resulting in chills, fever and general weakness. The pathogen primarily infect several wild rodents, and is transmitted later by insect bites. The pathogen reaches the humans by drinking contaminated water or through skin abrasions and the mucus membranes.

Campylobacter spp have been reported to cause intestinal infections. Though it spreads mainly through milk, the water borne sources are also considered important.

Mycobacterium is responsible for Tuberculosis. It can survive in water for several weeks. Though its transmission by water seems to be quite uncommon, yet it cannot be entirely ruled out (Mason, 1981).

Observations on estimation of wastewater pathogens and infective doses (Table 112) for human exposure, indicate that *Helminth, Salmonella* and *Mycobacterium* were very high in untreated wastewater while *Enterovirus* were very low. These pathogens were highly concentrated in both primary and secondary effluents. Lowest number of *Mycobacterium* were required, especially in childrens for the infective dose, while a comparatively high number of organisms of *Salmonella* and

Helminths for the infective dose. The critical oral uptake was 0.5 for *Mycobacteriurn* and 10-102 for *Salmonella.*

Table 112.Estimated concentration of wastewater pathogens and their infective dose.

Pathogens	Number of oorganisms/l			Infective dose	
	Untreated wastewater	Primary effluent	Secondary effluent	Number of organisms	Critical orel* uptake(l)
Salmonella	5.103	25.10	1.1.1 02	104-109	10-102
Mycobacterium	5.10	2.5.10-1	4.10o	1-9 (children)	0.5
Enterovirus	1.104	5.103	5.102	1-102	
Helminth ova	6.10	6.100	1.2.10-1	105-108	

* Based on typical hygienic conditions of the world.

(b) Viruses

Certain viruses that multiply in the alimentary tract (including the Oropharynx) of human beings, and may be excreted in considerable quantities in faeces, can be found in sewage and polluted waters but their mere presence is not necessarily evidence of significant risk to human beings. Sewage and polluted water contain several kinds of viruses, broadly including the following categories: (i) Enteroviruses (Polio-virus, Coxsackieviruses, Ecovirus), (ii) Adenovirus, (iii) Reoviruses, and (iv) Hepatitis viruses. Of these viruses Adenoviruses and Reoviruses are rarely transmitted through water; their main transmission takes place through respiratory route. Of the Adenoviruses the spread of Poliovirus by water has rarely if ever been demonstrated, because of the extremely high dilution of the virus and the consequent difficulty of isolating it, whilst the more direct faecal-oral route is the most likely mode of spread of the Echoviruses and Coxsackie-viruses (Chang, 1968).

Echoviruses and Coxsackieviruses cause Enterits in human beings, and outbreaks usually occur due to contamination of water by untreated sewage. The most concerned disease caused by water-borne viruses is however, infectious hepatitis (Jaundice). The hapatitis virus infects liver to induce haphazard production of excess bile, which gets entry into blood causing yellowness in the body.

Several epidemics of infections hepatitis have occurred all over the world, caused by polluted waters (Mosley, 1967; Taylor et al, 1966; Koff, 1970). A striking example was the epidemic of infectious hepatitis in Delhi in 1955-56, in which more than 2800 cases were indentified, with a case fatality rate of 0.9 per 1000 and an estimated total case incidence of 97,600 (Viswanathan, 1957).

Infectious hepatitis can also be spread by consumption of shellfish contaminated with sewage effluent (WHO, 1972).

(c) Parasites

The parasitic protozoan *Entamoeba histolytica* causes amoebic dysentery (intestinal Amoebiasis) in human beings. It infects large intestine causing ulceration which results in the release of necrotic mucous membrane and blood. The amoeba can also enter the portal circulation and cause hepatic abscess (liver necrosis). It is widespread throughout the warm countries (tropics and subtropics) of the world, and wherever sanitary conditions are poor. The amoebic cysts are resistant to chlorine doses given for water purification. Pine filtration alone is most effective to check both vegetative and encysted cells.

Several algal species have been found to be parasitic on many animals and macrophytes. Algae can produce dark spots on the scales of fish decreasing their economic value. A diano-flagellate (*Oodinium ocillatum*) infest gills of fish that may be fatal. Glenn et al (1960) observed an algae *Mucophilus cryprini* in the epithelium of the skin and gills, which increased the mortality rates. Other species reported to affect the fish are *Oodinium limneticum, Cladophora* and *Chlorochytrium*.

Some species of *Oodinium* can also infact annelids, worms and marine Tunicates. The algal taxa like *Chlorella, Zoochlorella, Carteria, Protococcus, Aphrydium, Coccomyxa, Scenedesmus, Chlamydomonas* and *Trockiscia* are common parasites of snails, muscles, water sponges, *Turbellaria, Hydra, Paramecium* and *Stentor*. Some algal species like *Chlorochytrium* can also infect the aquatic plants such as *Lemna, Elodea* and *Ceratophyllum*.

A few members of the group Nematoda (round worms) and Platyhelminths (Tapeworms and Flukes) are also human parasites, whose ova are released in urine and faeces in large numbers. They are usually transmitted by consuming sewage contaminated water, food and soil. The intermediate host for the tapeworm (*Taenia saginata*) is usually cattle, which can become infected by grazing in pastures with sewage sludge or by drinking sewage contaminated water. The encysted larvae of the tapeworm can be transmitted to human beings by water and contaminated food.

Some intestinal helminths, such as *Ascaris lumbricoides* and *Trichuris trichiura* usually spread by ingestion of contaminated soil, but they also become frequently waterborne. Distomatosis is another parasitic disease that mey be contracted by swallowing contaminated water containing cysts of species of *Fasciola* or *Dicrocoelium*.

Hydatidosis is a common disease of sheeps and dogs with cattle, pigs and other wildlife as intermediate hosts. Hydatid disease is a zoonosis that usually involves a dog-sheep-dog cycle in maintaining the reservoir of infection. It is occasionally transmitted to human beings through drinking water or food contaminated with the excreta of the primary hosts.

Dracontiasis, the infestation of *Dracunculus medinensis* (Guinea worms), a nematode parasite, can result from consuming contaminated water. From the patients intestine, the adult female migrates to the skin surface to deposit larvae, producing a cord-like thickening which ulcerates. The disease is common among the rural populations of many developing countries. This parasite is transmitted principally through open village wells and ponds infested with the Copepod intermediate host.

2. BIOLOGICAL AGENTS TRANSMITTED THROUGH WATER CONTACT OTHER THAN INGESTION

In economically advanced countries, direct contact with water, other than in personal hygiene, occurs mostly in recreational activities - swimming, waterskiing etc. and ordinarily involves only minimal hazards (Table 113). In many developing countries, on the otherhand, water in rivers, ponds, canals etc. is used for a variety of purposes - ablutions, washing of clothes, disposal of human excreta, domestic uses -so that these waters become highly polluted and serve as an important vehicle for the transmission of enteric infections, such as Cholera, Typhoid fever and the Dysenteries, and of certain parasitic infections.

Table 113. Human diseases capable of being transmitted through water contact other than ingestion.

Diseases	Cause	Agent
Internal Schistosomiasis	*Schistosoma mansoni* S. *japonicum*	Skin penetration of snails cercariae while bathing.
Genito-urinary schistosomiasis	*S. haemetobium*	
Swimmer's itch	-do-	Skin penetration of birds and rodents cercariae.
Leptcspirosis	*Leptospires* spp	Skin and mucous memb-rane penetration.

A few cases of paratyphoid fever have been reported to be caused by bathing in sewage polluted beaches. Some evidences are also there to acquire Mycotic diseases such as Thrush, and becoming infected with Enterovirus while bathing in the contaminated waters (WHO, 1972). Of the communicable diseases spread by the penetration, by parasites of the skin and certain mucous membranes, the most widespread is Schistosomiasis. Certain bacteria also cause disease in this way.

(a) Schistosomiasis and other communicable diseases

Schistosomiasis (also called Bilharzia) is a chronic, insidious, debilitating disease that may cause serious pathological lesions, saps energy, lowers resistance and reduces output of work. The disease is more prevalent in tropical regions, and it has been estimated that more than 200 million people suffer in the world over. In some endemic areas, the prevalence may exceed 50 percent. The accelerated construction of artificial lakes, reservoirs and water impoundments to meet the needs of agriculture and industry in developing countries must inevitably lead to an increase in freshwater snail populations and a corresponding increase in Schistosomiasis if preventive measures are not taken.

The Schistosomiasis disease in human beings is chiefly due to the species of blood flukes of the genus *Schistosoma*. The three most common species causing the disease are *S. mansoni*, *S. japonicum*, and *S. haematobium*. Of these three species, first two cause the intestinal manifestations, and the third is the causative agent of genito-urinary or vesical Schistosomiasis. The eggs, released in faeces or urine, hatch as miracidia reaching a free body of water, where they penetrate snails (such as *Biomphalaria*, *Bulinus* and *Oncomelania*). The snails are secondary hosts. They emerge from the snails in the form of cercariae ready to infect man. Penetration is through the skin, while wading, bathing etc. Water flowing with only a very gentle current, as in slow-moving canals, is preferred by all species of aquatic snails.

In many parts of the world, bathers in lakes may be affected by "Swimmer's itch". This dermatitis is caused by the penetration of the skin by cercariae of Schistosomea from animals, such as birds and rodents.

The incidence of Schistosomiasis is increasing in the areas where the artificial lakes, reservoirs and canals provide excellent habitats for the freshwater snails whose populations increase to much greater levels. One such striking example is the construction of Aswan dam in Egypt, which has led to a tremendous increase in the cases of Schistosomiasis because of the increase in the snail population of the area (Hammerton, 1972).

Other parasitic diseases where entry is through the skin are Ancylostomiasis and Strongyloidiasis. The Ancylostomiasis is caused by hookworms (*Necator americana* and *Ancylostoma duodenale*). The eggs passed in the faeces of the infected person, hatch and larvae emerge that develop into a filariform stage infective to human beings, by penetrating through the skin. Strongyloidiasis is caused by the nematode *Strongyloides stercoralis*, which inhabits the submucous tissue of the intestine. The mode of transmission is identical to that of Ancylostomiasis. While water may serve as the medium whereby these infective agents are swallowed, infection is usually acquired by skin penetration from soil, or by auto-infection. The symptoms of this abdominal disease are varying. Pulmonary symptoms may also develop with large number of larvae appearing in the sputum, once the larvae penetrate into the lungs.

Leptospirosis is the principal bacterial infection transmissible from vertobrate animals to human beings through direct contact with water. The natural hosts are wild and domestic animals that excrete Leptospires in urine, and human beings are infected through the skin and raucous membranes from contact with the water of contaminated ponds, canals, rivers etc.

(b) Biological agents in bathing beaches and coastal waters

With the increased use of bathing beaches and coastal waters for recreational purposes and because of the possible consumption of marine fish and shellfish from polluted waters, these habitats harbour biological agents which can cause epidemics in human beings. Poliomyelities among children living near the sea has been associated with superficial mycotic disease caused by Poliovirus, due to bathing in such waters.

There is no agreed basis for establishing a limiting bacteriological quality for coastal water, above which there is a danger of infection from bathing. Maximum acceptable counts in different countries for coliforms may be as high as 10,000 per litre of water; for *Streptococcus faecalis* the corresponding figure is 200 per litre (Brisou, 1968). There are no internationally accepted criteria for the quality of coastal water, with respect to microbial contamination.

(c) Water associated insect vectors

The most common and widespread disease caused by water associated insects is Malaria, which is produced by protozoan *Plasmodium*, and transmitted by an insect vector *Anopheles* mosquito. The other important disease transmitted by mosquito vectors are Filariasis, Encephalites, Yellow fever and Dangue fever (Table 114). The causal agents of the Filariasis are the protozoans, *Wuchereria bancrofti* and *Brugia malayi*, which are transmitted to human beings by *Culex pipiens fatiqans*. Sewage flowing in the open drains is an excellent breeding habitat for this insect in the urban areas. Filariasis affects more than 250 million people throughout the world, and continues to be one of the major parasitic infections.

Table 114. Diseases capable of transmission by water associated insect vectors.

Disease	Causative organism	Insect vector
Malaria	*Plasmodium* spp.	*Anopheles mosquitoes.*
River blindness	*Onchocera volvulus*	*Similium spp (blackflies)*
Yellow fever	*Arbovirus* group B	*Aedes mosquitoes.*
Sleeping sickness	*Trypnosoma gambiense*	*Tsetse* fly.
Filariasis	*Wucheria bancrofti, Brugia malayi*	*Cules pipens fatigans* mosquitoes.

Yellow fever is caused by a virus of Arboviruses group-B and spread by Aedes mosquitoes of subgenus *Stegomyia* and *Hoemagogus*. Other diseases caused by this group of virus are Dengue and Dengue haemorrhagic fever transmitted by the mosquito species of *Aedes aegypti* and *Aedes albopictus*. These species breed in the urban areas mostly in the open water containers and tanks, discarded tins, bottles and clay jars. The causative agents for viral encephalitis are also the viruses of Arbovirus group specific to different species of mosquito vectors of genera *Culex* and *Aedes*.

The species of the black flies of the genus *Simulium* transmit a helminth, *Onchocerca volvulus*, which spreads a disease called Onchocerciasis or river blindness. The vector usually breeds in clear springs running over natural rocky beds. The dams and concrete canals also provide suitable breeding grounds for them. The disease is of considerable health, social and economic importance in the vicinity of breeding grounds for the vector, where a large proportion of the resident population may become partially or completely blind.

Trypanosomiasis or sleeping sickness, which is widespread in Africa, is transmitted by tsetse flies of the genus *Glossina*, which breeds mostly in waterside vegetation. The vector carries the

Protozoan causal agents of the species of genus *Trypanosoma* (*T. rhodesiense* and *T. gambiense*) which are transmitted to human beings.

3. NUISANCE AQUATIC ORGANISMS

The organisms which directly do not cause any harm, but indirectly pose the problems of health hazards by making an otherwise safe water, unpalatable and unattractive, or interfering with the purification and distribution, can be termed as nuisance organisms. Such organisms include the biological slimes that accumulate on the interior surfaces of mains and upon which methane-utilizing bacteria may grow; algae and bryzoal growth such as *Plumatella,* which may interfere with the operation of filters; molluscs such as *Dreissena,* which may choke mains; Crustacea such as *Asellus* (water louse or snow bug), and nematodes, which while not in themselves pathogenic, may harbour bacetria and viruses in their intestines, thus protecting possible pathogens from destruction by chlorine; and certain algae, which may impart bad taste and odour to water. Table 115 lists the nuisance organisms normally found in drinking water.

Table 115. Nuisance organisms normally found in drinking water (WHO, 1972).

Organisms	Effects
Biological slimes	Choking of treatment plants and distribution system. Support for methane utilising bacteria. Risk of rendering water unaccptable.
Mollusca	Choking of water mains.
Plumetella, algae	Interference with filteration.
Asellus	Risk of rendering water unacceptable.
Nematodes	Possible concentration of pathogens.

The ubiquitous distribution of micro-organisms in water together with their unique combination of characteristics, allows to exert their effects in many places in a variety of ways. For example, micro-organisms can produce toxic secretions, taste and odours in water supplies. They can interfere with raw water purification, produce slime in waters, cause colouration, and corrode concrete and metal pipes. Among the wide array of micro-organisms, classified as nuisance organisms, algae are by far the most common and ubiquitous of all, which have been implicated in most problems of water supplies. However, the role of certain bacteria like *Siderocapsa, Sphaerotilus, Clonothrix, Leptothrix, Cremothrix, Caulobacter* and *Galionella* may also be important in clogging and production of tastes and odours in water. Some important nuisance algal species causing problems have been shown in Figure 38.

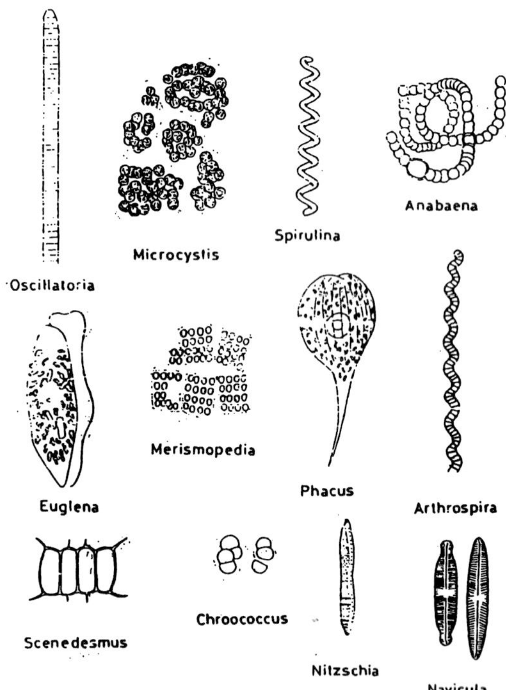

Fig. 38. Some important nuisance algal taxa commonly found in waters.

4. TASTE AND ODOUR PRODUCING ORGANISMS

Most of the problems associated with drinking water supplies, are with tastes and odours. The problems arise due to the presence of various algal species and certain other micro-organisms in the raw water. Several algal species secrete oily substances which impart particular taste and odour to the water. The odours can be developed due to formation of certain intermediate substances during the decomposition of aquatic vegetation by micro-organisms. Actinomycetes are also known to produce a variety of substances which give typical odour such as earthy or musty.

Palmer (1980) provided a comprehensive description and listed about 40 algal species commonly associated with taste and odour production. While most of these species belonged to Bacillariophyceae, Cyanophyceae and pigmented flagellates, a few also belonged to Chlorophyceae including desmids (Table 116). The typical odours produced by algae are fishy, grassy, musty or earthy, aromatic, medicaland septic. The aromatic odours resemble to that of some specific flowers and vegetables such as geranium, nasturtium, muskmelon and cucumber.

Odour and taste are the prime considerations for the potable water. Natural water depending upon their impurities have specific odour and taste. Peaty, silica and chalky waters have earthy odour. Decaying water weeds like *Chara*, rotten hay and straw impart an odour like that of decaying fish. Contamination with sewage effluents may give the odour of Hydrogen sulphide. Fungi growing on decaying plant material yields a musty odour. Chlorinated wasters with phenol traces give very strong Chlorophenol odour. The odours produced by algae depend upon their concentration in the water body, that is, moderate and abundant (Table 117).

Table 116. Some important algal species associated with tastes and odours in water (after Palmer, 1980).

Blue-green algae	Diatoms
Anabaena circinalis	*Asterionella gracillima*
Anabaenaa planctonica	*Cyclotella comta*
Microcystis cyanea	*Diatoma vulgare*
Aphanizomenon flos-aque	*Fragilaria construens*
Cylindrospermum musicola	*Stephanodiscus niagarae*
Gomphosphaeria lacustris	*Synedra ulna*
Oscillatoria curviceps	
Rivularia haematites	*Tabellaria fenestrata*
Symploca muscorum	**Flagellates**
Nonmotile green algae	*Ceratium hirundinella*
Chara vulgaris	*Chlamydomonas globosa*
Chladophora insigins	Chrysosphaerella longispina
Cosmarium partianum	*Cryptomonas erosa*
Dictyosphaeriwn ehrenbergianum	*Dinobryon divergens*
Gloeocystis planctonica	*Euglena sanguinea*
Hydrodictyon reticulatum	*Glenodiniuni palustre*
Nitella gracilis	*Mallomonas caudata*
Pediastrum tetras	*Pandorina morum*
Scencdesmus adundans	*Peridinium cinctum*
Spirogyra majuscula	*Synura uvella*
Staurastrum paradoxum	*Uroglenopsis americana*
	Volvox aureus

Table 117. Odours produced by Algae

Species	Moderate	Abundant
Anabaena	Glassy, Nasturtium	Septic
Aterionella	Geranium, spicy	Fishy
Dinobryon	Violet	Fishy
Melosira	Geranium	Musty
Oscillatoria	Grassy	Musty
Stephariodiscus	Geranium	Fishy
Synura	Cucumber	Fishy
Tabellaria	Geranium	Fishy
Volvox	Fishy	Fishy

The taste produced by algae are difficult to be separated from odours, and are catagorised as sweet or bitter. In some cases, sour taste can also be recognised associated, especially with acid or septic odours.

5. COLOUR CAUSING ORGANISMS

Colour in water means those hues inherent within water itself which result from colloidal substances and materials in solution. In natural waters colour may occur due to the presence of humic acids, fulvic acids, metallic ions, suspended matter, Phytoplankton and industrial effluents. Algae imparts green colour to water while water with excess of silt appears brownish. Organic matter and iron impart a yellow hue to the water.

The presence of several algal species, especially those with small cells can impart colouration to waters. This is more pronounced in uncovered water storage tanks. During storage of water in utensils, tubs and buckets, coloured marginal rings can appear due to algae. The main colours produced by algae include yellow-green, green, blue-green, red or brown to black. The important algae implicated in colouration of waters are *Anacystis, Ceratium, Clamydomonas, Chlorella, Cosmarium, Euglena orientalis, E. rubra, E. sanguina, Oscillatoria prolifera* and *O. rubescens*.

6. SLIME PRODUCING ORGANISMS

Several species of algae and bacteria produce slime or mucilaginous matter. The production of slime in waters may be undesirable to certain industries like food, paper and pulp. The industries using heat transfer systems may also be affected due to slimy growth in the heat exchangers, declining their efficiency. Many of the blue-green algae frequently produce slime. Some important genera recognised as slime producers are *Aphenocapsa, Gloeocapsa, Batrachospermum, Chaetophora, Cymbella, Gloeotrichia, Gomphonema, Oscillatoria, Palmella, Phormidium* and *Spirogyra*.

7. FILTER AND SCREEN CLOGGING ORGANISMS

During the purification of raw water, both slow and rapid sand filters are frequently clogged with the growth of algae. The most common algae which often clog the filters, belong to diatoms such as *Asterionella, Cyclotella, Fragilaria, Diatoma, Melosira, Navicula, Nitzschia, Stephanodiscus, Synedra* and *Tabellaria*. The individual cells of the algae *Tabellaria* are joined in a *zigzag* chain, united with gelatinous cushion, giving a flexible structure more effective in clogging the filters. *Fragillaria* has a tendency to form flakes or network of strands that cause much faster clogging. Moreover, even after the death of diatoms, the siliceous structures remain intact capable of plugging the pores in filters.

The members of some other algal groups also play an important role in clogging the filters. These include *Anabaena, Gloeotrichia, Oscillatoria, Phormidium, Rivularia (Cyanophyceae), Chlorella, Cladophora, Closterium, Hydrodictyon, Mougeotia, Palmella, Spirogyra, Tribonema, Ulothrix, Zygnema* (Chlorophyceae) and pigmented flagellates such as *Ceratium, Dinobryon* and *Trachelomonas* (Chrysophyceae and Euglenophyceae). The *Palmella* has a thick mucilaginous mass around the cells, and *Tribonema* forms a net like structure clogging the filters much faster.

8. CORROSION CAUSING ORGANISMS

The inner surfaces of the pipes through which water is supplied, is not sterile but harbours a thin membrane of biological organisms - bacteria, fungi and actinomycetes. Algal species can also grow in pipes and tanks where sunlight reaches for the growth. This biological growth gradually starts corroding the material of pipes making the water dirty (Collingwood, 1979). Oborn and Higginson (1954) observed that the submerged concrete became pitted and friable due to growth of the green and blue-green algae along with the mosses and lichens. The metals can be corroded by the release of oxygen from algae. Oxygen combines with the protective films over the metal surface making them prone to corrosion. The algae can also bring about certain other physico-chemical changes in water, such as pH, Carbon dioxide, Calcium carbonate, that have a direct bearing on the rate of corrosion.

Some important corrosion causing algae are *Chroococcus, Chaetophora, Diatoma, Euglena, Phormidium* and *Protococcus*, which corrode concrete. *Oscillatoria* is capable of corroding the steel.

9. TOXIC ORGANISMS

A large number of both marine and freshwater algae secrete toxic substances in water. Most of these toxic species are the members of blue-green algae. The blue-green algae are ingested in drinking water by animals, with the algae dying in the digestive tracts and releasing their toxins. There appears to be three basic types of following toxins:

(1) *Anabaena flos-aquae* contains "anatoxin A,B,C and D". Anatoxin A is the best studied of the four and is an alkaloid (nitrogen containing compound of low molecular weight).

(2) *Microcystis aerugenosa* contains a powerful toxin called "microcystin" which is composed of a polypeptide of 10 aminoacids. The minimum dose necessary for death is 0.5 mg of toxin per kilogram of body weight.

(3) *Aphanizomenon flos-aquae* contains a mixture of "saxitoxin" and three related toxins. The saxitoxin is the same alkaloid responsible for paralytic shellfish poisoning and has a minimum lethal dose of 10 mg per kg of body weight. Saxitoxin selectively blocks the flow of Sodium ions through excitable membranes, thereby stopping nerve coductance, an action that results in respiratory paralysis. In addition to causing the death of stock animals, the toxin in moderate quantities in drinking water cause diarrhoea in human beings.

Heaney (1971) indicated that large quantities of algae if swallowed orally, can be lethal. This may be the cause of cattle death in drought ridden areas of tropics where the *Microcystis* bloom commonly develops in the village ponds. Other species which affect the cattle in the same way are *M. flos-aquae, M. toxina, Anabaena circinalis, A. lemmermanni, Nodularia spumigena, Aphanizomenon flos-aque, Gloeotrichia enchinulata, G. pisum* and *Rivularia fluitans*.

Heise (1951) reported that certain allergies like contact dermatitis and hay fever can be caused by algae such as *Anabaena*,

Microcystis and *Lyngbya contorta*. Dillenberg and Dehnel (1960) reported that toxic blue-green algae can cause severe headache, high fever, nausea, vomiting, gastrointestinal disorders, exhaustion, pain in muscles and joints.

In marine habitats, a flagellate (*Gonyaulax*) produces a toxin that cause severe illness in human beings after consuming the clams who have fed upon this algae. The blooms of *Gonyaulax, Gymnodium* and some other dianoflagellates usually occur near the shores. The algae *Gymnodium veneficum* has been found to be lethal to certain fishes, arthropodes and echinoderms. There are evidences that *Lyngbya majaoula, L. aestuari* and *Prymnesium parvum* cause fish kill in marine areas.

Control of aquatic biopollutants

The diverse enteric water-borne pathogens and nuisance organisms can be controlled as under:

Control of water-borne pathogens

Enteric pathogens in dometic sewage can be inactivated or destroyed by giving a proper treatment to sewage that also reduces the organic matter and other chemical substances present in it.

Carrington (1978) has asserted that the processes of secondary sewage treatment are able to inactivate several pathogenic organisms. Activated sludge can destroy the *Salmonella* spp to the extent of 70-99 percent; viruses 80-99 percent and protozoa and metazoa 0-99 percent. Trickling filters, on the other hand, remove these organisms to the extent of 70-99, 80-99 and 0.99 percent respectively.

There is a necessitity to purify raw water into a clean and bacteriologically safe water before it can be supplied to the people. In the absence of the purified water supply, some methods at the individual level can also be employed to get the safe water. These include from simple boiling of water to disinfection by use of ultra violet radiation, chlorine and iodine.

Control of nuisance algae

The algal growth in the bodies of water can be controlled by a number of ways depending upon the problems and kinds of algae. The use of algicides is one of the popular methods for algal control. A large number of chemical compounds, both organic and inorganic are known to function as algicides. Some of the common algicides are Copper sulphate, Potassium permanganate. Chlorine and its derivatives, rosin amine, guinones, substituted hydrocarbons, quaternary Ammonium compounds, Amide derivatives and phenols.

Copper sulphate is most common algicide in use, mainly because of its greater economic feasibility and relative nontoxicity at the doses ordinarily employed. It has been found to be extremely powerful even at low doses to several nuisance algae like *Anabaena, Anacystis, Aphanizomenon, Gomphosphaeria, Rivularia, Asterionella, Cyclotella, Fragilaria, Melosira, Hydrodictyon, Oedogonium, Rhizoclonium, Spirogyra, Ulothrix Dinobryon, Synura, Uroglenopsis* and *Volvox*. However, some species are highly resistant and cannot be killed by Copper sulphate at the safe doses. These are *Calothrix, Symploca, Ankinstrodesmus, Chara, Coelastrum, Elakatothrix, Kirchneriella, Nitella, Pithophora, Scenedesmus, Tetrastrum, Eudorina* and *Pandorina*.

The minmum doses of Copper sulphate which can be effective against algae may vary greatly depending upon the temperature, alkalinity, organic matter content of water, and the abundance of algae, besides some other minor factors. Successful control of algae has been reported in Croton lake, New York with a continuous dose of Copper sulphate at 0.18 mg/l for almost 20 years (Palmer, 1980).

Pithophora and *Ulothrix* can be effectively controlled by rosin amine D acetate (RADA).

Mechanical removal may be considered important in certain situations, such as in case of algae which is periphytic or form dense filamentous masses, or where the use of algicide is not feasible.

In small reservoirs, even filtration of the whole water through the rapid sand filter has been attempted to control algae (Leibee and Smith, 1953). The periodicle increase in natural turbidity of rivers and lakes is the natural phenomenon which kills the algae by cutting light. But this is only temporary as the algae reoccur again with the subsiding of turbidity. To check the algal growth in natural bodies of water, proper watershed management is also an important aspect that prevents the entry of nutrients in waters.

The algal control in the treatment units is carried out by the processes of coagulation, sedimentation and filtration that also removes other suspended impurities. These can remove more than 90 percent of the algae under good operational conditions. The chlorination, which is usually given to control bacteria in water, also kills algae. Prechloronation helps to remove several algae, especially motile ones, during the process of coagulation. The use of microstrainer has also been made in removal of algae from the water in treatment units. In case of the walls and other surfaces of the treatment units get heavily coated with algae, scrapping is a good method for their removal, which can be carried out after emptying the water from the unit. Control of algae in the distribution system and the storage tanks can be best achieved by the use of algicides and covering the open tanks to cut the light.

Chapter-9

EUTROPHICATION

The term *eutrophication* has been derived from a Greek word *eutrophos* meaning corpulent or rich or well nourished. The term was first used in ecology in connection with the remnants of the extinct lakes rather than live ones. The concept of eutrophication was first introduced by Weber (1907) to describe the nutrient conditions determining the flora of north German peat bogs. He found that the upper layers have more nutrients in comparison to the lower ones, as the original lakes received much higher nutrients supply prior to their transformation into bogs. He used the terms *eutrophic* (rich in nutreints) and *oligotrophic* (poor in nutrients) to distinguish between these two layers. He described the species requiring higher concentration of essential elements as *eutrophent* and those surviving at low nutrient concentration as *oligotrophent*. Naumann (1919) employed these terms to describe the Swedish water types and their Phytoplankton as *oligotrophic, mesotrophic* and *eutrophic*, according to the concentration of phosphorus combined nitrogen and calcium along with the associated density of photoplankton population. He, however, did not specify the ranges of chemical concentration for distinguishing the lake types. Thienmann (1918) in his investigation of the German lakes suggested the deeper lakes as oligotrophic and the shallower ones as eutrophic. His conclusion was based on the fact that hypolimnion water of deeper lakes retained most of its oxygen when dead plankton and detritus fell into the bottom waters because of the large volume of hypolimnic water per unit amount of organic matter falling into it as compared to the shallower lakes whose hypolimnic water is too less to supply enough oxygen. Pearsall (1921) in his classical study of English lakes expressed oligotrophy to eutrophy as unevolved to evolved lakes. He opined that eutrophic nature was primarily due to erosion in the drainage basin which deposit lesser or greater amount of silt and is the source of potassium and phosphorus. Storm (1930) also followed these criteria to describe the lakes. However, he emphasized upon the ratio of water volume to sediment surface and stated that the potential amount of nutrient that can diffuse from mud to water will be more in shallow water since the ratio of water volume to sediment surface is less. It is accepted by all, that for eutrophication, the supply of food should be by autotrophic means, and where the main supply of organic matter is by allochthomas means, the lakes are called *dystrophic* having often low primary productivity. However, the lakes with dystrophic conditions but having higher productivity are called mixotrophic (Jernefelt, 1925).

Eutrophication refers to the natural or artificial addition of nutrients such as nitrogen and phosphorus in a form which increases the productivity of water and brings consequent changes in plant and animal life in the waterbody perhaps reducing its utility and beauty and threatening its very existance in the course of time. When the effects of eutrophication are undesirable as when the value of a lake as a source of water supply and recreation is diminished, eutriphication is considered as a form of pollution (Sharma, 1986).

Whatever the mode of origin, right from the moment of its birth, fertility and productivity of a lake system begin to increase. This maturity process is known as eutrophication. In reservoirs, this ecological succession is through three stages, that is : (a) Initial fertility, (b) Trophic depression, and (c) Final fertility. Lakes may exhibit different trends during their life-history depending upon different controlling factors (Figure 39).

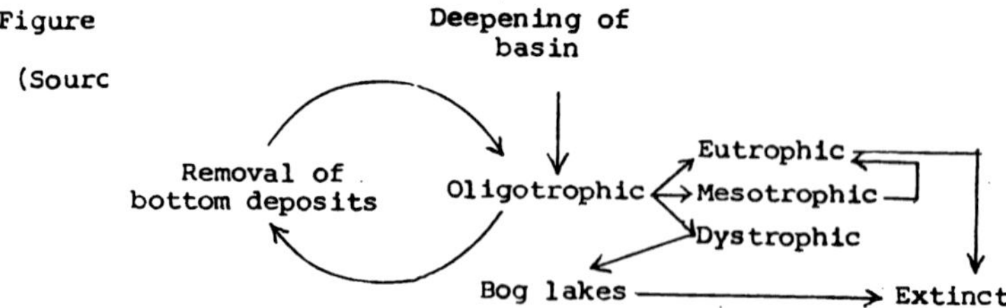

Figure 39. Development of lakes.

Eutrophication refers to excessive fertilization of a water body which results in the nuisance growth of the aquatic plant material such as algae and macrophytes. This in turn leads to water quality deterioration, taste and odour problems, oxygen depletion, decline of fisheries, the clogging of water ways; in some cases, excessive loss of water through evaporation-transpiration. For these reasons utilization of water bodies becomes more difficult, and at times even impossible. The uses which are most effected are drinking water supply, tourism, supply of water for irrigation and industries, recreational activities, sports and commercial fisheries and navigation. Cases are also known of eutrophication creating difficulties in hydroelectric power plants and increased corrosion of man made structures in water. Eutrophication in fresh water can cause serious intoxication in cattles. Excessive supply of phosphorus and nitrogen compounds in the form of municipal and agricultural sewage has been identified as the primary culprit causing eutrophication.

By combining the concept of trophy as seen by the above mentioned founders of Limnology one may deduce that eutrophication is the process of aging of the lakes which manifest in the following three directions:

(i) Siltation.

(ii) Accumulation of excessive nutrient such as combined nitrogen, phosphorus and minerals.

(iii) Manifestation of progressively increasing primary production.

Various physico-chemical and biological characteristics have been used to calculate the indices of eutrophication (Table 118). All these indices suggest that the lake is of eutrophic nature.

Table 118. Indices of eutrophication.

Index	Value	Trophic state	Author
Calcium	14.57	Eutrophic	Ohle,1934
Monovalent:Divalent	0.5	Eutrophic	Zafar,1959
Primary productivity ($g/m^3/yr$)	100.40	Eutrophic	Rodhe,1969
Myxophycean	4.06	Eutrophic	Nygaard,1949
Chlorophycean	0.58	Eutrophic	-do-
Euglenophyta	0.2	Eutrophic	-do-
Compound	11.54	Eutrophic	-do-

EUTROPHICATION PROCESS

The eutrophication process is basically a natural phenomenon which gets accelerated by increased nutrient supply through human activities. The process of eutrophication starts as soon as the lakes are formed because of the entry of nutrients by natural means, but the rate of eutrophication remains quite low under natural compositions. For the sake of convenience, the process of eutrophication have been discussed under natural and accelerated processes, though its basic features remain essentially the same.

(a) Natural eutrophication

The lakes generally originate as oligotrophic and have only limited quantities of nutrients depending upon the mode of their formation and composition of original sediments. These nutrients are insufficient to produce any significant algal growth. At this stage the lakes have only autochthonous nutrients (indigenous nutrients cycling therein), which usually recycle completely decomposed after death. As the Allochthonous nutrients (nutrients from outside) start entering the lake, the process of eutrophication sets in. The principal natural sources of nutrients are the natural run-off, fall of leaves and twigs from the surrounding vegetation, periodical submergence of the nearby terrestrial vegetation, rain-fall and bird-dropping etc.

The build up of nutrients through this slow mode of entry gradually starts increasing the growth of algae. When the algae die and decompose, the locked nutrients are again made available to the fresh algal growth. During each cycle, the nutrients are progressively increased in the water body. With the advancement of eutrophication, the cycling of nutrients are unable to maintain an equilibrium between production and decomposition with the result that an ever-increasing organic matter is introduced in the lakes which ultimately gets deposited at the bottom. The thickness of the bottom sediments increases slowly with time, leading to the formation of swamps, bogs, marshes and finally to the extinction of the water body.

The process of eutrophication have been shown in Figure 40, which indicates that a lake passes from oligotrophic to eutrophlic condition through some arbitrary intermediate stages called oligo-mesotrophic, mesotrophic and mesueutrophic. It is also evident that with the progress of eutrophication an increasing quantity of nutrients come in circulation, and cycles become unable to complete. The speed of eutrophication does not depend only on the rate of nutrient supply, but other factors like climate and morphometric features also becone important. The tropical or hot climate usually supports a higher rate of eutrophication as it favours higher nutrient utilization and algal growth in comparison to cold and temperate climates.

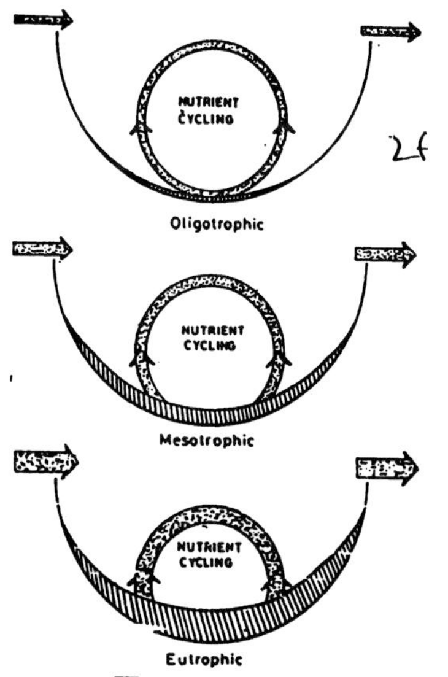

Figure 40. The process of eutrophication showing the evolution of lakes.

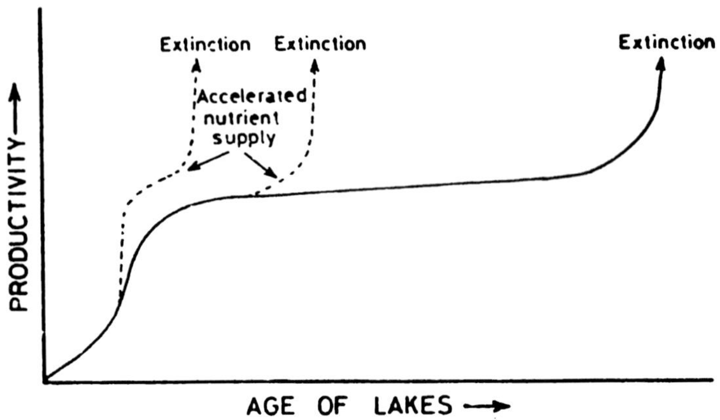

Figure 41. The reaction between the rate of eutrophication (productivity) and age of lakes.

(b) Accelerated eutrophication

The process of eutrophication is greatly augmented by the increased supply of nutrients through various human activities such as discharge of domestic sewage, industrial wastes, agricultural and urban run-off. Increased levels of air pollution also make the water bodies rich in nutrients through their transport with rain or by dry fallout. This increased supply of nutrients triggers the algal growth at a much faster rate, thus increasing the speed of eutrophication, which otherwise would have been a slow natural phenomenon. The relation of the rate of eutrophication with the aging of lakes have been shown in Figure 41. It shows that though all lakes mature in due course by eutrophication and become extinct, extinction time is considerably reduced by accelerated eutrophication. The process of eutrophication is therefore, sometimes referred to as aging of lakes. It is also evident that the rate of eutrophication slows down with time due to reduced light penetration as a result of increased turbidity and consequent fall in primary production.

CAUSES OF EUTROPHICATION

The causes of eutrophication in lakes in and around Udaipur are as follows:

1. The lakes are gradually shrinking in both depth as well as peripheral dimensions due to deposition of silt, on account of undesirable human activities both in the catchment area and around the lake margins.

2. All along the banks large number of human beings take bath and wash their clothes. In addition, the Washerman activities also add a large amount of phosphate load to the lake water.

3. Another important cause of eutrophication of lakes in the disposal of untreated sewage. A large number of points through which such city sewage is thrown into the lakes can be noticed.

4. The local inhabitants residing along the lake margin throw all sorts of garbage and kitchen waste into lakes.

5. Agricultural practices in the catchment area as well as banks and beds of the lakes results in the washing in of agrochemicals. There are several synthetic compounds which are sprayed on land and water crops to kill pests. These are not completely consumed by target species and often pass on in relatively unbroken and lethal condition to other trophic level organisms causing highly injurious effects.

6. On account of construction of new colonies as well as hotels, the vegetation of hillocks around the lakes has been removed and rocks are exposed. Due to this a large amount of silt in addition to domestic sewage from the residential colony and hotels fall into the lakes.

7. The other conspicuous features threatening the lakes is that the people living around the lakes have no sanitation facilities. They directly come to the lake and sit on the foot-steps of the lakes. As a result, a large amount of human excreta gets accumulated and ultimately falls in the lake water (Vyas, 1989). The main disadvantages of eutrophication are:

(a) Abundance of Cyanophycean bloom.

(b) Filling of lake basin with organic material exerting an increase in oxygen demand on the overlying waters.

(c) Developing taste and odour problem affecting the recreational activities.

Vyas (1986) used Nygaard's four trophic state indices to test the eutrophication trends in various lakes in and around Udaipur (Table 119). High values of various Nygaard indices as evident from the observations suggest that the lakes around Udaipur are eutrophic. Values for the Nygaard's compound index for trophic state and annual production indicate that Swaroop sagar and Rangsagar are in the worst stage of pollution. The effects to these lakes are severe because of their smallness and closeness to the town.

Table 119. Nygaard's indices of eutrophication for Udaipur lakes (Vyas, 1986).

Index	Rang sagar	Fateh sagar	Swaroop sagar	Pichhola lake	Udai sagar
Myxophycean	7.28	3.03	2.14	–	5.68
Chlorophycean	0.91	1.5	0.51	–	0.79
Euglenophycean	0.10	0.009	-	–	–
Compound	17.57	11.42	15.65	12.17	11.54

The prominent water bodies of Bhilwara (Gandhisagar and Gwadia) have been assessed in respect of the degree of eutrophication. Various sources of the nutrient salts in these lakes have been enlisted in Table 120.

Table 120. Sources of nutrients in the two lakes of Bhilwara (Ranga and Yadav, 1989).

Lake	Sources of nutrients
Gandhi sagar	Sewage wastes, detergents, cattle and human excreta, agricultural run-off and wastes from Dalda factory and process houses, algal blooms.
Gwadia	Agricultural run-off, waste of Bhilwara process houses. Cattle excreta, algal blooms.

It has been observed that sewage wastes, fertilizer run-off from agriculture fields, effluents from Dalda factory process houses are dumped into Gandhi sagar through a stream. Fertilizer run-off from agriculture fields, cattle excreta and effluent of Bhilwara processing houses indirectly come into Gwadia. Process houses are using a variety of nearly 40 chemicals, which are partly discharged in the effluents. These effluents are rich in nitrogen and phosphorus. The eutrophication of Bhilwara lakes is due to progressive addition of these nutrients.

The causes of eutrophication may be ascribed to the impact of human interference. Some of the more important of them are as follows:

(a) Siltation

Deforestation in the catchment area accelerates soil erosion leading to increased siltation in the lakes. In addition to this quarrying of stones and new settlements (residential as well as commercial) on the hillocks around the lake (where once thick wooded vegetation occurred) are also increasing siltation (Vyas et al, 1981; 1982).

(b) Accumulation of nutrients

Nitrate and phosphate fertilizers are used in huge quantities to increase yield. A part of these inorganic salts which are not utilized by crops are washed into the lakes through streams during the rainy season. There are several synthetic compounds which are sprayed on land and aquatic crops to kill the pests. These are not completely consumed by target species and often pass on in relatively unbroken and lethal condition to other trophic level organisms causing highly injurious effects (Vyas et al,1981)

(c) Sewage

All the human settlements and hotels situated in and around lakes discharge their waste water and sewage directly into the lake. According to Sylvester (1961) the domestic sewage is one of the important sources for the increase in the nitrate content, and human and domestic cattle excreta for the phosphate. Presence of large amount of chlorides in the water of the lakes is suggestive of pollution by organic matter chiefly of animal origin (Adoni, 1975). In such sewage polluted waters heterotrophic members become dominant. In the absence of oxygen, anaerobic process occur in which fermentation of carbohydrates and sugar, and the putrification of protein take place. In such a break-down of proteins, hydrogen sulphide is produced which bubbles out giving fould smell (Vyas et al, 1981; 1982; Agarwal, 1982).

(d) Detergents

Unlike soaps, detergents are not easily decomposed by bacteria and are richer source of phosphates that hasten the process of eutrophication (Vyas et al, 1981; 1982).

According to Prescott(1939) the differences in the dominant Phytoplankton assemblage of lakes reflects their trophic levels. He reported that oligotrophic lakes are characterised by Chlorophycean flora with a conspicuous desmid element. The eutrophic lakes were found to be dominated by Myxophycean forms. Rawson (1956) also reported the same type of algal assemblage in two types of lakes in Canada. The eutrophic lakes had mainly Cyanophycean element dominated by *Microcystis*.

The effects of eutrophication can be traced not only by the study of Phytoplankton but also by their production potentialities. Vinberg (1955) considered lakes with annual production upto 30 gC/m^3/year as oligotrophic, 30-200 gC/m^3/year as mesotrophic and more than 300 gC/m^3/year as highly eutrophic. Sreenivasan (1963) observed that Yarcaud lake with high production and bloom of *Microcystis aeruginosa*, and classified it as eutrophic lake.

According to Hutchinson (1969), eutrophication is a natural process which literally means "well nourished or enriched". It is a natural state in many lakes and ponds which have a rich supply of nutrients, and it also occurs as part of the aging process in lakes, as nutrients accumulate through natural succession. Eutrophication becomes excessive, however, when abnormally high amounts of nutrients from sewage, fertilizers, animal wastes and detergents enter streams and lakes, causing excessive growth or bloom of micro-organisms and aquatic vegetation.

Eutrophication thus specifies the change (increase) in biological productivity and nutrient content which all lakes and reservoirs undergo during their life. The problems that arise following the discharge of wastes containing nutrient salts into water bodies come not from the nutrients themselves but from the changes they induce in the aquatic productivity and the Phytoplankton. The term eutrophication denotes the lowering or deterioration of water quality for domestic, recreational and other uses. The continuous deposition of organic debris and its degradation by benthic bacteria etc in oligotrophic reservoirs is generally so slow that no significant algal and higher plant populations can be supported, and the water remains "clean". On the other hand, in a eutrophic pond, the deposition of materials is so great that a variety of nutrients becomes released into the water which can support large populations of algae and higher plants. Oligotrophic waters tend to have a nitrogen : phosphorus ratio greater than 10. This ratio approaches 10 as some sewage is discaharged into an oligotrophic lake. In the normal course in nature the transformation of an oligotrophic lake to eutrophic one, is a very slow process, occurring in thousands or millions of years. However, human intervention leads to a remarkable acceleration of this rate with the result that a completely oligotrophic lake often becomes eutrophic within the course of a few decades. Eutrophication is, in fact, just another form of pollution.

Eutrophication concept has been confused since the early 1920's because it was based on such instantaneous parameters as oxygen tension, nutreint concentration and Phytoplankton density. This kind of empirical and static concept neither supported any functional relationship between the different state variables nor the prediction of the dynamics of the systems response to parturbations. Later, the concept of rates, that is, the velocity of exchange between various mass and energy compartments, was introduced and it then became possible to relate concentrations to supply and consumption as well as Phytoplankton density to primary production. The classical trophic state indices are connected to external parameters by the rate of primary production. Quantitatively, eutrophication can be simply defined as an increase in the primary productivity of a lake (Kumar, 1977).

INDICATORS OF EUTROPHICATION

Of the nutrients entering the lake, a major proportion becomes incorporated into algae which release the nutrients back into water either when still alive or after their death and decay. In fact, certain blue-green algae can serve to indicate eutrophic conditions. *Oscillatoria rubescens, Aphanizomenon flosaquae, Microcystis aeruginosa* have often been regarded as indicators of eutrophication. Eutrophic waters are rich in plankton, have larger number of individuals but these belong to very few species, and water blooms occurr frequently. The common algae of such waters belong to Cyanophyta, Bacillariophyta (*Melosira, Fragilaria, Asterionella*) (Kumar, 1977).

Eutrophication is commonly accompanied by a concomitant change in the species composition of the aquatic biota. These changes may be caused either directly due to the nutrients, or more commonly, may be caused by some underlying homeostatic factors, for example, change in grazing behaviour of animals. Such changes in homeostatic relations can in certain cases decay or prevent the required desirable changes or improvements in a lake even after the nutrient load has been diminished significantly. Another similar factor that can cause delayed effects is the deposition of nutrients in bottom sediments. Such nutrients may continue to diffuse into and enrich the water long after further inputs of nutrients into the lake have been checked (Kumar, 1977).

With increasing eutrophication, the diversity of the Phytoplankton community of a lake decreases and the lake finally becomes dominated by blue-green algae. Some striking examples of blue-green algal lakes are the hypertrophic Wolderwijd and Veluwemeer lakes in the Netherland. In these lakes, *Oscillatoria agardhii* is the dominant alga. The natural population of this species are successively limited by phosphorus, light and combined nitrogen.

Deep lakes are generally more sensitive to hyper-eutrophication as compared to shallow lakes (Mur, 1980). The biomass concentration in deep lakes is low and the blue-green algal species are all gas vacuole forms that grow in distinct strata. In contrast, in shallow lakes stratification does not occur and the light intensity becomes low only when the numbers of algae become very large. Non gas vacuolate species tend to dominate here.

Different species and groups of species are characteristic of different nutrient regimes and this is one cause for the difference in the vegetation of the upper and lower reaches of the same river and in streams on different rock types. A group of macrophytes characteristic of eutrophic conditions are *Nuphar luteum, Schoenoplectus lacustris, Rorippa amphibia, Sparganium emersum, Sagittaria Sagittifolia* and *Enteromorpha*. Occurrence of the following species however, does not change the nutrient status of indicators by the presence of above species *Lemna minor, Spagganium erectum* and *Potamogeton pectinatus*. The presence of *Alisma-plantago aquatica, Potamogeton lucens, P. perfoliatus, Elodea canadensis, Phalaris arundinacea, Ranunculus epp, Polygonum amphibium* and fringing herbs indicates that the habitat is less eutrophic or possibly cleaner (Varshney, 1981).

EFFECTS OF EUTROPHICATION

Eutrophication effects the physico-chemical parameters and the biological parameters of the water body:

(a) Physico-chemical effects

Water pollution can be considered as a departure from the balance between photosynthesis and respiration. At equilibrium (P=R), the chemical and biological composition of water remains unchanged, a stage that mostly occur only in non-polluted water, with no external supply of nutrients. An eutrophic water body is one where photosynthesis exceeds the respiration. It is characterised by a progressive accumulation of algae which ultimately leads to an organic overloading. In deep stratified lakes, an excessive production at the surface of the lake is paralleled by saprophytic conditions at the bottom. The bacterial decomposition at the bottom of lakes releases the nutrients from the dead organic matter. When respiration exceeds photosynthesis, dissolved oxygen gets rapidly exhausted forcing reduction of several oxidized chemical compounds like NO^{-}_{3}, SO^{-2}_{4} and CO_2 into N_2, NH^{+}_{4}, H_2S CH_4. which are harmful to several aquatic species and produce typical odours. LC_{50} for H_2S for some aquatic organisms has been reported to be as low as 11 mg/L (Poole et al, 1978).

Figure 42 indicates the depth profile of dissolved oxygen and nutrients in oligotrophic and eutrophic water bodies. In oligotrophic waters the concentration of dissolved oxygen and nutrients do not vary much with depth, but typically the surface waters of eutrophic water bodies have higher oxygen and lower nutrients in comparison to the bottom layers. The eutrophication induces many other physico-chemical changes in the water. Increase in photosynthetic rate leads to the consumption of greater quantities of bicarbonates resulting in the formation of more and more carbonates raising the pH of water.

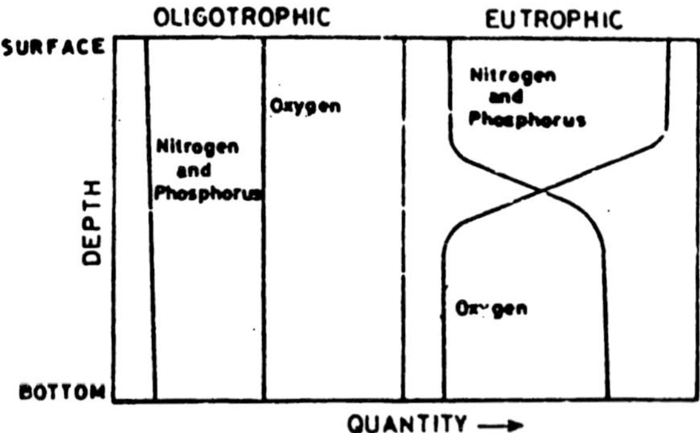

Figure 42. Depth profile of dissolved oxygen and nutrients in oligotrophic and eütrophic lakes.

The carbonates formed in this manner may get precipitated as their Calcium and Magnesium salts at high pH. Substantial quantities of phosphorus may also precipitate along with these carbonates (Wetzel, 1968).

The eutrophic water bodies show typical diel cycles of dissolved oxygen and Carbon dioxide, as the concentration of these gases is regulated by photosynthetic activity, a diurnal phenomenon. During day time, there is an increase of oxygen and decrease of Carbon dioxide, while the trend is reversed at night.

During the winter season when photosynthetic activity subsides, the pH is decreased owing to the excessive accumulation of free Carbon dioxide and at this time the previously precipitated carbonates may get redissolved. The other adsorbed minerals like phosphates and ammonia can also be released from the sediments as the redox potential falls due to decrease in oxygen at the bottom (Mortimer, 1941, 1942; Pillos and Swanson, 1975). However, if phosphorus is once precipitated to sediments, it cannot be liberated again even in highly oxygen-tense conditions (Schindler et al, 1973). The constant entry of phosphorus and its rapid exchange between abiotic and biotic components keep the pace of eutrophication.

(b) Biological effects

Almost all physico-chemical changes in water bring about a change in flora and fauna due to homeostatic factors which leads to the attainment of new equilibrium. In doing so, many desirable species including fish are replaced by undesirable species. There is an algal succession resulting in the dominance of blue green algae which have very low nutrition value in the food chain, and many of them produce blooms. Some important bloom forming blue-green algal genera include *Microcystis, Anabaena, Oscillatoria* and *Aphanizomenon*. Algae from other groups like *Chlorella, Scenedesmus*, and *Kirchneriella* can also form blooms. Filamentous green algae such as *Spirogyra, Cladophora and Zygnema* form a dense floating mat or blanket on the surface when the density of the bloom becomes sufficient to reduce the intensity of solar radiation below the surface. These blankets often give shelter to several undesirable insects and mosquitoes. Many algal species have allelopathic effects on other algae which frequently results in the formation of typical associations and assemblages of algae in eutrophic waters. The relative abundance of various algal groups in relation to increasing eutrophication have been shown in Figure 43.

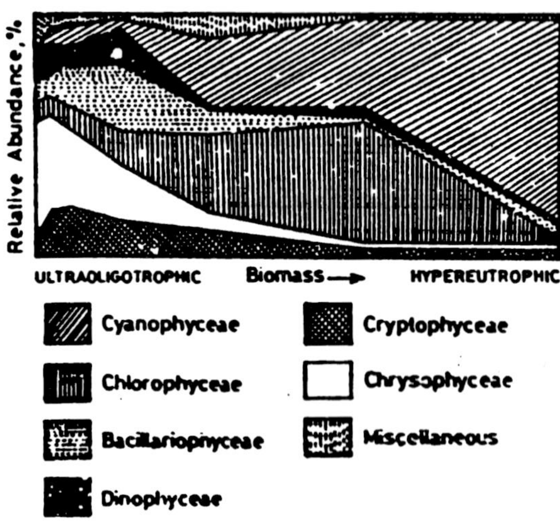

Figure 43. The relative abundance of various algal groups with the advancement of eutrophication.

Nutrient enrichment has very limited direct effect on zooplankton communities, but indirect effect may be significant. The diversity of zooplankton remains high if the diversity of Phytoplankton is also high, as often found in oligotrophic or moderately nutrient enriched waters. In oligotrophic waters, the size of Phytoplankton is usually smaller than 50 µm, but the size of zooplankton enlarges making the Phytoplankton communities to be controlled by zooplankton grazing. On the contrary, increasing eutrophication makes the size of Phytoplankton larger, and zooplankton smaller, so that the Phytoplankton communities can no longer be controlled by the latter.

As the changes occur in water due to eutrophication, the characteristics of sediments also change. There is an accumulation of organic matter which affects the benthic communities. They become dominated by Chironomids and Polycheates with fewer number of Mullusks, Crustacea and Insects.

Eutrophication of moderate level may be beneficial to fish production as it increases the food supply for fish in the form of algal growth. Fish ponds are often fertilized with nutrients to accelerate the algal growth for increasing fish productivity. However, with the increase in the level of eutrophication, dominance of algal groups is taken over by blue-green algae making the edible or game fish to be replaced by hardy species of very little economic value.

Algal blooms are very unsighty and react with recreational persuits. When algae die and decompose, foul odours and tastes develop in water. The scum of algae may act as a barrier to the oxygen penetration into water resulting in fish kill.

Initially oxygen is overproduced, but when the blooms die, it is stripped from the water. The rate of oxygen consumption of the dead algae is much higher than that by respiration of live algae. The fall in the level of oxygen causes mass mortality of fish and many other organisms which require comparatively higher oxygen concentration. The minimum level of oxygen required for the protection of aquatic life at different temperatures have been shown in Table 121.

Table 121. Minimum dissolved oxygen levels required for the protection of aquatic life (USEPA, 1972).

Temperature (°C)	100% Saturated oxygen levels (mg/L)	Minimum level of oxygen required for aquatic life protection	
		DO(mg/L)	% saturation
1.5	14	6.8	48.6
7.7	12	6.8	56.7
16.0	10	6.5	65.0
21.0	9	6.2	68.9
27.5	8	5.8	72.5
36.0	7	5.8	82.9

Some blooms release toxic substances into the water which are capable of killing fish and other aquatic species. Toxic substances produced by *Microcystis aeruginosa* and *M. flos aquae* are found to be toxic to human beings and cattle.

The algal blooms cause discolouration of water and attract water fowl which further contribute to the pollution of water. The overall effect make the water much less suitable for our various needs particularly recreation, fish production, and domestic use.

CRITICAL NUTRIENTS IN EUTROPHICATION

Nutritional requirement of algae consists of about twenty elements, and absence of any of them can restrict the growth. Vallentyne (1973) has shown the relative requirements of these nutrients and their availability in the water (Table 122). Oxygen and hydrogen are readily available from water, but other nutrients may be in short supply. These data further indicate that phosphorus remains in greatest short supply for plant growth followed by nitrogen, carbon and silicon, as their demand far exceeds the quantities naturally occurring in water. However, these data are generalised, and in any given situation either of these nutrients or their combinations may become limiting.

Table 122. Relative quantities of essential elements required by plants and their natural supply in river water (Vallentime, 1973)

Element	Plant Requirement (%)	Present in Water (%)	Demand/Supply ratio
1	2	3	4
Oxygen	80.5	89.0	1
Hydrogen	9.7	11.0	1
Carbon	6.5	0.0012	5000
Silicon	1.3	0.00065	2000
Nitrogen	0.7	0.000023	30000
Calcium	0.4	0.0015	< 1000
Potassium	0.3	0.00023	1300
Phosphorus	0.08	0.000001	80000

1	2	3	4
Magnesium	0.07	0.000	< 1000
Sulphur	0.06	0.0004	< 1000
Chlorine	0.06	0.0008	< 1000
Sodium	0.04	0.0006	< 1000
Iron	0.02	0.00007	< 1000
Manganese	0.0007	0.0000015	< 1000
Boron	0.001	0.00001	< 1000
Zinc	0.0003	0.000001	< 1000
Copper	0.0001	0.000001	< 1000
Molybdenum	0.00005	0.0000003	< 1000
Cobalt	0.000002	0.000000005	< 1000

It is now a well accepted fact that in most situations phosphorus and nitrogen frequently limit the algal productivity and biomass. Phosphorus is considered to be the most important of all in limiting eutrophication in almost all situations. The importance of nitrogen is secondary to phosphorus as it can also be fixed by nitrogen-fixing blue-green algae which can thrive well in nitrogen deficient water, where as there is no supplementary source of phosphorus. The nitrogen-fixing algae can convert inert N_2 gas into its assimilative nitrate form. The nitrate ions, while get easily leached out from the soil, phosphate is more tightly bound to the soil particles, and cannot be easily entrained in run-off water. As a result, nitrogen can enter the water more frequently through the natural sources than phosphorus. Nitrogen, however, can become limiting in the tropical or hot climate due to increased denitrification. Nitrogen in many cases may also become secondarily limiting for algal growth, especially at the time of plentiful phosphorus supply.

It has been observed that the lower limits for nitrogen and phosphorus which can restrict the algal growth are 0.2 ppm and 0.05 ppm respectively. Any increase in their concentration shall spurt the algal productivity. However, it is also observed that if the concentration of these two elements increases beyond 20 ppm algal growth is suppressed.

Kuentzel (1969) and Kerr et al (1970) suggested that Carbon dioxide from bacterial decomposition of organic matter in the main factor in stimulating obnoxious algal blooms. Many scientists felt that Carbon dioxide may be a limiting factor for algal growth in eutrophication, but it was clearly established later that Carbon dioxide cannot become critical for the growth as it can readily diffuse from the atmosphere at the time of need. In most cases, the fraction of carbon derived from the atmospheric Carbon dioxide in the algal biomass may range from 50-95 percent. Tailing et al (1973) indicated that upto 95 percent of the annual input of carbon may be supplied by the atmosphere in lakes.

Schindler and Fee (1974) conducted experiments on artificial nutrient enrichment in lakes to see the relative importance of nitrogen, phosphorus and carbon in promoting algal growth. In one experiment, a lake was split into two portions by a curtain: one half was fertilized by carbon and nitrogen, and the other half with carbon, nitrogen and phosphorus. The half of the lake that received phosphorus developed algal blooms, while the other half did not. The recovery of the lake in their experiments

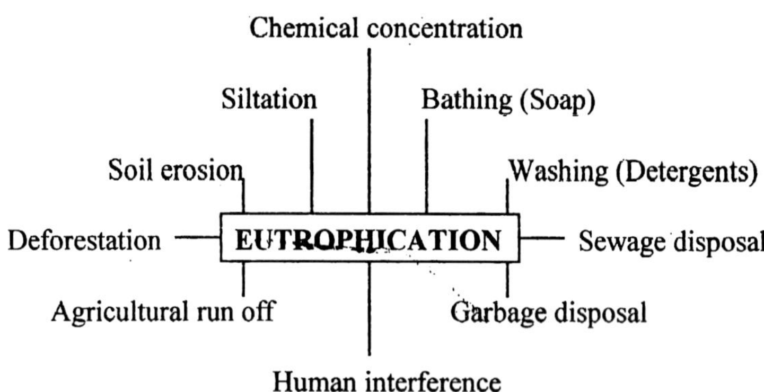

Fig 44 : Causes of eutrophication

after termination of phosphorus addition, but still continuing enrichment with nitrofen and carbon for further three years, confirm the role of phosphorus in eutrophication. These authors suggested that phosphorus control is an efficient primary step in preventing the eutrophication problems.

In certain cases, the algal growth in water can also be limited by a few other growth factors like Vitamin-12, thiamine and biotin and some trace nutrients like manganese, zinc, molybdenum, vanadium, Boron and cobalt. But under ordinary circumstances, the trace elements, required by algae are already present in natural water, while the vitamins and other growth factors can be supplied by natural run-off, bottom sediments and by metabolic activities of other organisms.

While many workers have observed about one or the other nutrient or substance as the limiting factor, it remains most likely that the multiplicity of interrelated nutrient factors ultimately determine the lake productivity in a given situation.

FACTORS OTHER THAN NUTRIENTS

Although nutrients are the prime cause of eutrophication, they should not be considered in isolation, because algal growth may also be influenced by other factors. These include mainly climatic, morphometric nature of water bodies, geographical features and physico-chemical characteristics of water (Figure 44). These factors, which influence the utilization of nutrients, must be considered before evaluating the extent of eutrophication.

Temperature of water is an important climatic factor which strongly influences the algal growth. A large number of algae require relatively higher optimum temperatures. The rate of eutrophication is usually higher in tropical waters than in temperate waters. Seasonal temperature changes cause the lakes to overturn which results in mixing of the whole water column. Early spring blooms of certain algae are the result of this overturn when the nutrient released at the bottom by decomposition of organic matter are made available to the surface Phytoplankton. This process is sometimes referred to as *internal fertilization*.

Light is another important factor which limits photosynthesis in water. The light penetration in water can be related to the presence of turbidity and colour that restrict the depth of photic zone. The latitudinal differences in photosynthesis rate can partly be ascribed to the differences in the solar insolation. Wind causes the upper layers of water to mix patchiness, caused by light winds, may increase productivity in certain pockets of the lakes.

Mean depth of water body is an important factor which influences the rate of eutrophication. Shallow waters are more susceptible to eutrophication in comparison to deeper waters, as they are well mixed and receive light in the greater part of the water mass. The rate of internal fertilization is also higher in shallow waters since the released nutrients can easily get distributed to the upper layers because of their well mixed nature, where as most deep lakes are stratified during considerable part of the year. The increased shore line and littoral area provide a greater shallowness to the water body. Index of lake permanence (volume at full capacity/shoreline at full capacity) is a good index of shallowness. A decrease in this index will accelerate the rate of eutrophication. The nutrient input from catchment area and direct rain-fall is proportional to the surface area of lake, but nutrient concentration in water is related to the volume of water. Thus, the quantities of nutrients entering per unit surface area will provide higher concentration in water in shallow lakes than the deeper lakes. Schindler (1971) observed that an increase in the productivity and biomass occurs when the volume of water decreases in relation to the catchment area and lake surface area.

Overall accumulation of nutrients in a lake is dependent on the retention period of water. With fall in retention period of water, the nutrients are increasingly flushed out from the system which reduce their progressive accumulation in the water bodies.

Nutrient availability is also affected by the physico-chemical conditions of water. For example, organic colour may influence primary production and algal species by changing pH, chelating trace metals nutrients and by absorbing light. Dissolved oxygen may affect the availability of nutrients especially in relation to sediment regeneration and leaching.

SOURCES OF NUTRIENTS

Water bodies may be enriched with nutrients through both natural and man-made sources, nevertheless, their quantities may greately differ from source to source. The man-made sources are much more significant contributors of nutreints than the natural sources. A brief account of some important sources of nutrients is given below:

Rainfall and Atmospheric Deposition

Rain water may contain varying amounts of nutrients depending upon the local atmospheric pollution. Rain water on an average, contains 0.16 to 1.06 mg/l of nitrate nitrogen, 0.04 to 1.7 mg/l of ammonia nitrogen and from traces to 0.1 mg/l of phosphorus (Carrol, 1962). Fruh and Lee (1968) indicated that approximately 41000 kg of nitrogen enters each year with rain water in lake Mendota in United States. The percentage of the total loadings attributable to atmospheric deposition in Lake Superior have been estimated to range from 60 percent for nitrogen, 21 percent for phosphorus and 40 percent for lead.

Urban and Rural Run-off

The run-off water entrains significant quantities of nutrients and organic matter from the soil and other surfaces. Urban run-off contains storm water drainage with organic and inorganic debris from various surfaces both paved and grassed, and fertilizers from gardens and lawns. Rural run-off originates from sparsely populated areas with little or no land devoted to agriculture. Eroded soil may also be entrained in this run-off. Average characteristics and nutrient loadings of urban and rural run-offs have been shown in Table 123 and 124.

Table 123. Average characteristics and nutrient loading of urban run-off water (Weibel et al, 1964).

Parameter	Concentration mg/L	Mass discharge kg/acre/year
BOD	18	15
Suspended solids	77	331
Tot Kjeldahl nitrogen	1.2	4.1
Total phosphorus	0.3	1.1

Table 124. Average nitrogen and phosphorus in rural run-off (Fruh, 1968).

Origin of run-off	Total nitrogen kg/acre/year	Total phosphorus kg/acre/year
Forest run-off	0.6-1.4	0.13-0.36
Surface irrigation return flow	1.1-10.8	0.42-1.76
Sub-surface irrigation return flow	17.2-753	1.1-7.3

AGRICULTURAL RUN-OFF

The enrichment material in the agricultural ron-off is derived from the fertilizers applied to the crops, and from farm, animal houses. Nitrogen used as fertilizers may get converted into nitric acid in soil solubilizing calcium, potassium and other ions which become highly liable to leaching.

DOMESTIC SEWAGE

Sewage is the common source of nutrient and organic matter and undoubtedly the greatest contributor to the eutrophication. Large quantities of nitrogen and phosphorus are excreted by humans and animals which get their way into sewage. On an average 2 gm of phosphate phosphorus per day in released through urine and faeces by an average person. Phosphate detergents in sewage are also important contributor of phosphorus. Conventionally, treated sewage (without tertiary treatment) may contain 15 to 35 mg/l of total nitrogen and 6-12 mg/l of phosphorus (Hume and Gunnerson, 1962). Untreated sewage, besides nutrients, also adds large quantities of nitrogenous organic matter.

INDUSTRIAL WASTES

The nutrients in industrial effluents are variable in quality and quantity depending upon the processes and type of industry. The wastes from certain industries, particularly fertilizers, chemicals and food are rich in nitrogen and phosphorus.

WATER FOWL

The droppings of water fowl is a source of nutrients which may cause the local problem of eutrophication, especially in small bodies of water. The overall effect of this source on the whole water body may be negligible. It has been estimated that wild ducks contribute 5.8 kg of nitrogen/acre/ year and 2.55 kg of total phosphorus/acre/year to the lakes.

GROUND WATER

Ground water is some cases may act as a source of nitrogen to the surface waters. It is, however, not a recognised source at all places, but may be an important factor in certain areas. It has been estimated that about 42 percent of nitrogen in Wisconsin surface waters comes from ground water (Hasler, 1968).

Besides the above factors, the land use pattern and human activities in the catchment or watershed area also influence the enrichment of water bodies. Removal of vegetational cover may have two-fold effects. First, it leaves greater surface area to erode allowing increased leaching of nutrients, and *second*, permits more erosion of soil to increase the sediment volume of lakes. Phosphorus generally remain tightly bound to the soil particles, difficult to be leached out by run-off, but may be transported to the water bodies along with the mass of eroded soil.

MEASUREMENT OF EUTROPHICATION

Eutrophy or the eutrophic state of a water body is a condition destined by a combined effect of various biological and physico-chemical factors. As all the parameters do not have equal importance, one cannot measure the eutrophication with a single parameter. However, a logical approach is to take multiplicity of parameters with their relative importances to develop certain indices, which can accurately reflect the level of eutrophy in waters. Some important parameters often used to indicate eutrophication are listed in Table 125.

Table 125. Important parameters for indicating level of eutrophication in water bodies (D=the values decrease, I= the values increase with the progress of eutrophication).

Physical	Chemical	Biological
Transparency (D)	Nutrient level (I)	Algal bloom frequency (I)
Mean Depth (D)	Chlorophyll a (I)	Algal species diversity (D)
	Conductivity (I)	Littoral vegetation (I)
	Dissolved solids (I)	Diversity (D)
	Hypolimnetic O_2 (D)	Zooplankton (I)
	Epilimnetic O_2 supersaturation (I)	Bottom fauna biomass (I)
		Bottom fauna diversity (D)
		Primary Production (I)

Trophic state or the levels of eutrophication are generally viewed as a continuum with no interim break points. However, the systems developed so far demarcate this continuum by a few arbitrary boundaries like oligotrophic, oligomeso-trophic, mesotrophic, eutrophic and so on, perhaps for the sake of convenience. Lack of a precise definition of eutrophication and our poor knowledge of understanding of the relationship between various physico-chemical and biological parameters, make it difficult to assess eutrophication. The frequent result of this is that the different indicators give conflicting values for the trophic state of a lake. Beeton (1969) observed that the biological indicators indicate eutrophic nature of lake Ontario, but the physico-chemical parameters indicate oligotrophic conditions.

Quantitatively, oligotrophic lakes are low in nutrient, support small algal populations, have low primary production and high transparency with greater mean depths. On the other hand, eutrophic lakes generally have high quantities of nutrients, high densities and biomass of algal populations with high primary production, low transparencies and usually shallow. However, no precise and universally accepted quantification of these data has been made that reflect the true picture of the level of eutrophication.

One can check the result of chemical analysis by using standard solutions, but there exists no such analogous standard for measuring trophic structure. The validity of the results can only be judged on a subjective basis to see if the characterization of water bodies appear to be reasonable to measure the trophic state.

Some important systems of measuring the trophic state have been discussed below:

NEWTON AND FETTEROLF'S SYSTEM

This is a relative system of indices, where the water bodies are classified only with respect to each other. The system was developed by Newton and Fetterolf (1966) by taking into consideration some chemical indicators of trophic condition. The equation gives a composite score for any given water body. The increase in this score is proportional to the increase in the level of eutrophication.

$$\text{Score} = 2\,(OP) + 0.1\,(COD) + (ON) + (NH_3)$$

where,

OP = soluble orthophosphate ($mgPO_4.\,L^{-1}$)
COD = chemical oxygen demand ($mg.\,L^{-1}$)
ON = organic nitrogen ($mgN.\,L^{-1}$)
NH_3 = free ammonia ($mgN.\,L^{-1}$)

The lakes can be compared for their level of eutrophication by comparing the score.

LUESCHOW ET AL SYSTEM

Lueschow et al (1970) also developed a relative system by taking into consideration the following seven parameters:

(i) Hypolimnetic dissolved oxygen
(ii) Plankton density
(iii) Secchi disc transparency
(iv) Organic nitrogen
(v) Total inorganic nitrogen
(vi) Soluble phosphorus
(vii) Total phosphorus

The average on the monthly values of these parameters are taken into consideration except oxygen whose minimum value is used.

Ranking is made for the waterbodies with regard to one parameter at a time. This is done for all the parameters separately. points are assigned to each waterbody equivalent to its position in the ranking, that is, point 1 for first, 2 for second, 3 for third and so forth. Points are then totalled for the seven parameters to get a complete score for each lake. Final trophic ranking of the lakes is done on the basis of the composite score.

THOMAS'S SCHEME

Thomas (1969) suggested a scheme in the form of a sequence diagram of the nitrogen and phosphorus present in minimum quantities in relation to the trophic state of a water body, when it is transformed from oligotrophic to eutrophic state (Figure 45).

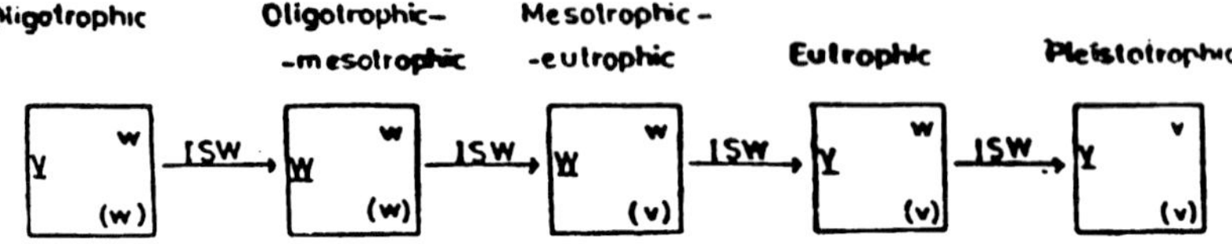

Fig. 45. The relative quantities of nitrogen and phosphorus with the increase in eutrophication (Thomas. 1969).

ISW = Incoming sewage water

V = more than 0.2 mg/L of NO_3

W = less than 0.2 mg/L of NO_3 (line left)

V = more than 20 mg/m^3 of PO_4

W = less than 20 mg/m^3 of PO_4

(v) = more than 40 mg/m^3 of PO_4 after ignition

(w) = less than 40 mg/m^3 of PO_4 after ignition

SHANNON AND BREZONIC'S SYSTEM

Shannon and Brezonic (1972) developed trophic states indices (TSI) in the form of a score by using mathematical equations for clear and coloured waterbodies:

For Clear Lakes

TSI = 0.936 [1/SD] +0.827 [COND] +0.907 [TON] +0.748 [TP] + 0.938 [PP] + 0.892 [CHA] + 0.579 [1/CR] + 4.76

For Coloured Lakes

TSI = 0.848 [1/SD] + 0.809 [COND] + 0.887 [TON] + 0.768 [TP] + 0.930 [PP] + 0.780 [CHA] + 0.893 [1/CR] + 9.33

Where,

SD = Secchi disk transparency (metres)
COND = Specific Conductance ($\mu S.cm^{-1}$)
TON = Total organic nitrogen ($mg.L^{-1}$)
TP = Total phosphorus ($mg.L^{-1}$)
PP = Primary Productivity ($mg\ c.m^{-3}\ h^{-1}$)
CHA = Chlorophyll ($mg.\ m^{-3}$)
CR = Pearson's cation ratio

The "ISI" score can be used to divide the lake in five trophic categories as follows.

TSI > 10 = hypereutrophic
TSI 7-10 = eutrophic
TSI 3-7 = mesotrophic
TSI 2-3 = oligotrophic
TSI 2 = ultraoligotrophic

WETZEL'S SCHEME

Wetzel (1975) proposed a scheme for trophic classification using the concentration of nitrogen and phosphorus present in the water (Table 126).

Table 126. Trophic classification of water bodies on the basis of nitrogen and phosphorus concentration in water (Wetzel, 1975).

Trophic Category	Inorganic-N mg/m^3	Total phosphorus mg/m^3
Ultraoligotrophic	< 200	< 5
Oligo-mesotrophic	200-400	5-10
Meso-eutrophic	300-650	10-30
Eutrophic	500-1500	30-100
Hypereutrophic	> 1500	> 100

RODHE'S SCHEME

Rodhe (1970) have the approximate ranges of Phytoplankton production in oligotrophic and eutrophic lakes which can be used for trophic categorization of lakes (Table 127).

Table 127. Approximate ranges of Phytoplankton primary production in oligotrophic and eutrophic lakes (Rhode, 1970).

Lakes	Mean rates in growing season (mgC.m^{-2}.d^{-1})	Annual rates (g C.m^{-2} yr^{-1})
Oligotrophic	30-100	7-25
Eutrophic		
a. Natural	300-1000	75-250
b. Polluted	1500-3000	350-700

ARCEIVALA'S SCHEME

Arceivala (1981) used some biological parameters and primary production to demarkate the trophic categories of lakes (Table 128).

Table 128. Ranges of some biological parameters and Phytoplankton production in the lakes of different trophic status (Arceivala,1981).

Parameter	Oligotrophic	Mesotrophic	Eiarophic
1. Algal cells density (per ml)	0-2000	2000-20000	> 20000
2. Algal biomass (mg. L^{-1})	0-1	1-10	> 10
3. Chlorophyll a peak values in photic zone (mg. m^{-3})	0-3	3-20	>20
4. Primary production			
(a) Annual rates (g $C.m^{-2}$ yr^{-1})	10-50	50-200	200-> 750
(b) Mean daily rates in growing season (g $C.m^{-2}$ d^{-1})	0.03-0.2	0.2-1	> 1

NYGAARD' S SCHEME

Nygaard (1949) proposed, on the basis of occurrence of few algal groups, five algal indices to evaluate the trophy of the waters. The calculation of these indices have been shown in Table 129. The number of genera pertaining to each group is put in the calculation.

Table 129. Calculation of Nygaard's trophic indices.

Index	Calculation	Oligotrophic	Eutrophic
1	2	3	4
Myxophycean	$\frac{\text{Cyanophyceae}}{\text{Desmideae}}$	0.00-0.04	0.1-3:0
Chlorophycean	$\frac{\text{Chlorococcales}}{\text{Desmideae}}$	0.00-0.7	-0.2-9.0
Diatom	$\frac{\text{Centric diatoms}}{\text{Pennate diatoms}}$	0.0-0.3	0.0-1.75

1	2	3	4
Euglenophycean	Euglenophyceae / Cyanophyceae + Chlorococcales	0.0-0.2	0.0-1.0
Compound	Cyanophyceae + Chlorococcales + Centric diatoms + Euglenophyceae / Dedmideae	0.01-1.0	1.2-2.5

UTTORMARK AND WALL'S SCHEME

Uttermark and wall (1975) have developed a lake condition index by using penalty points for characteristics like dissolved oxygen, transparency, fish kill and recreational use impairment. The penalty points are calculated as follows:

Table 130. Lake condition index calculated by using penalty points.

A.	Dissolved Oxygen	Penalty points	
		Max. depth < 30 ft	Max. depth > 30ft
1.	Dissolved Oxygen in hypolimnion greater than 5 $mg.L^{-1}$ at virtually all times.	0	0
2.	Concentration in hypolimnion less than 5 $mg.L^{-1}$ but greater than 0 $mg.L^{-1}$	1	2
3.	Portions of hypolimnion void of oxygen at times	3	4
4.	Entire hypolimnion anoxic	5	6

REVERSIBILITY OF EUTROPHICATION

Considerable attention is now being given to stop or reverse the eutrophication process with a view to ensure an adequate supply of clean water for drinking and other purposes, and also with a view to prevent the lakes from too premature a death. The following procedures have been recommended by various limnologists to slow down the eutrophication process:

(1) Limiting the amount of nutrient entering the lake.

(2) Reduction in the amount of nutrients solubilized in water through microbial decomposition of bottom sediments.

(3) Harvesting and removal of algal blooms and mechanical removal of higher plants; this can reduce the amount of nutreints recycled into the water upon death of algae and higher plants.

(4) Removal of dissolved nutrients from water chemically or physically.

(5) By encouraging natural food-webs, which can remove the algae and subsequently harvesting the fish etc.

(6) By controlling the growth and multiplication of algae and higher plants through application of appropriate doses of copper sulphate and sodium arsenite respectively.

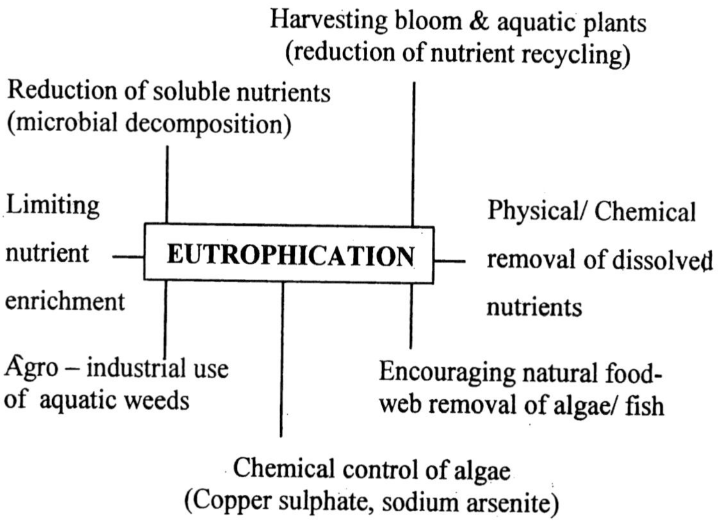

Harvesting bloom & aquatic plants (reduction of nutrient recycling)

Reduction of soluble nutrients (microbial decomposition)

Limiting nutrient enrichment

EUTROPHICATION

Physical/ Chemical removal of dissolved nutrients

Agro – industrial use of aquatic weeds

Encouraging natural food-web removal of algae/ fish

Chemical control of algae (Copper sulphate, sodium arsenite)

Figure 46 : Reversibility of Eutrophication

Of the above methods of controlling eutrophication, that involving limitation of nutrient input is certainly the best and the most reliable.

Tertiary treatment of wastewater effluents is now being increasingly employed to decrease the phosphate content of the effluent before being discharged into lakes and rivers.

"Microbial intervention" technique envisages the management of the water body in such a manner as to selectively favour the decomposer microbes over the algae. Since the bacteria and algae compete with each other for the nutrients derived from decaying lake sediments, artificially stimulating bacterial multiplication will lead to a disruption of the algal food-web. In other words, the 'self cleansing' ability of the reservoirs water will be maximized through microbial activity (Erickson and Reynolds, 1971).

Eutrophication can also be reversed or controlled by removal of nutrients such as phosphorus and nitrogen compounds from the water. Phosphorus can be removed by pre-precipitation, simultaneous precipitation, and post-precipitation (Baalsrud and Balner, 1973). Zirconium oxychloride (100 ppm) has been found to precipitate as much as about 95 percent of phosphate content in algal culture media (Kumar and Rai, 1978), Zirconium oxychloride does not exert any adverse effects on fish. Nitrogen can be removed by (a) biological nitrification and denitrification, (b) air stripping of ammonia from an alkalized wastewater, (c) ion exchange, (d) electrodialysis, and (e) reverse-osmosis. Out of these methods the first two are most commonly employed.

To stop and reverse the eutrophication process following measures can be taken:

(A) Measures outside the water body:

(1) Treatment of waste water (elimination of phosphorus and nitrogen).

(2) Complete diversion of the waste waters.

(3) Primary sedimentation basins.

(4) Direct precipitation of nutritive substances in the affluents.

(5) Water-shade protection (reforestation).

(B) Corrective measures inside the water body:

(1) Harvesting and removal of algal blooms and higher plants (biological manipulation).

(2) Physical manipulation.

(3) Chemical and sedimentary manipulation (nutrient precipitation).

(4) By encouraging the setting up of natural food-webs (e.g. daphnids and fishes) which can remove the algae and subsequently harvesting the fish etc.

Chapter-10

WATER QUALITY STANDARDS

The term *water quality* is intimately related to water pollution. Polluted water is water which has more negative qualities than it has positive ones. Water quality refers to the physical, chemical and biological characteristics of water. The physical characteristics of water include temperature, clarity and similar qualities. Chemical water characteristics include the presence, and amount if present of organic and inorganic substances in solution, and the way that these substances are bonded or dispersed in water. Biological water quality characteristics include identity, impact and organisms which are present in water. In simple words, polluted water, is water which has been abused, defied in some way, so that it is no longer fit for some use. Water quality can be described in terms of various physical, chemical and biological parameters.

Physical parameters include colour, odour, temperature, solids, oils and grease. Colour can be defined relative to type and density, the type being related to whether it is true colour (dissolved) or apparent colour (filterable). Odour is described by type and threshold odour number, which is related to the odour free water required for diluting an odorous water sample to a non-odorous level. Total solids are comprised of suspended solids and dissolved solids, and each of these fractions can be further divided into organic (volatile) and inorganic (fixed) components. Turbidity is another measure of the solids contents, and it is related to light transmittance through water. Settleable solids describe the materials present in solution that will settle by gravity in a one hour period. Specific conductance (conductivity) is a measure of the inorganic dissolved solids present in ionic form. In surface water oil and grease is of interest relative to nuisance considerations.

Chemical parameters can be subdivided into organic and inorganic constituents. Several tests can be employed to describe the organic characteristics of water. The most used test is the biochemical oxygen demand (BOD), which is defined as the amount of oxygen required by bacteria in decomposing organic materials in a sample under aerobic conditions at 20°C over a 5 day incubation period. Other tests that describe the organic content of water include the chemical oxygen demand (COD), total organic carbon and total oxygen demand. Inorganic parameters of potential interest in water quality characterization include salinity, hardness, pH, acidity, alkalinity and the content of iron, manganese, chlorides, sulphates, heavy metals (mercury, lead, chromium, copper, zinc etc.), nitrogen (organic, ammonia, nitrate, nitrite) and phosphorus. Salinity and chloride content are a measure of the salt in water. Hardness is caused primarily by divalent metallic cations that have soap consuming potential, the major ones being calcium and magnesium. Nitrogen and phosphorus contents are of interest due to their nutrient characteristics.

Bacteriological parameters include coliform, fecal coliforms, specific pathogens and viruses. Total coliforms and fecal coliforms are used as indicators of the presence of pathogens. Specific pathogens such as *Salmonella* may be relevant for certain specific waters. Minimal information is available on virus types and concentrations in surface waters (Canter, 1977).

The quality of usable water on the earth is extremely small, any waste of this water by polluting it will further reduce the available supply.

Polluted water is hardly of any use for most purposes. It cannot be utilized for drinking purposes because of its ingerent health risk. Water with high salt content is not suitable for agriculture and for most industries. The quality of water also interferes with the aesthetic and economic persuits of water bodies by affecting the fish and other biological organisms. However, the water which is not suitable for drinking purposes may be good for irrigation, or water unsuitable for irrigation may be quite suitable for industrial cooling or fish farming. Thus, it can be seen that each use of water has its own limits or the degree of pollution that can be accepted. Every use of water requires a certain minimum quality with regard to the presence of dissolved and suspended materials of both chemical and biological nature. This minimum quality of water should ensure no harm to the user.

The achievement of this minimum quality of water for diverse uses has led to the formulation of water quality criteria, water quality objectives and water quality standards.

Water quality criteria can be considered as specific requirement on which a decision or judgement to support a particular use will be based. The criteria for various uses are developed from the experimental data, and our current knowledge of the health, ecological and economic effects of water quality. The criteria levels for drinking water source have to be based upon removability of the constituents at the water treatment plants and available data on the human health.

The criteria for the waste water generated from industries may be based on the nature of the industry and effects of their constituents on the water and land. Criteria are not a set of static values but are subject to modifications as the scientific data gets updated and more and more knowledge is gathered.

Water quality objectives can be defined as an aim or goal with regard to the water quality which is to be achieved. It is not a rigid and authoritative value as a standard, and does not have the enforcement ekement of requirements.

Water quality standards applies to any definite measure established by an authority by limiting concentrations of constituents in water which ensure the safe use of water and safeguard the environment. The fact that it has been established by an authority makes standards somewhat rigid, official or quasi-legal. Sometimes standards may not be fair and based on sound scientific knowledge, as it is possible that they might have been established somewhat arbitrarily on the basis of inadequate technical data tempered by a cautious factor of safety. Due to these facts, the standards may change with the accumulation of more scientific knowledge.

To attain the desired water quality objectives, the standards can be applied in two ways. One type, called *effluent standards* are applicable to the quality of municipal, agricultural or industrial waste discharge into water sources and on the land. The other type of standards are concerned with the water receiving bodies. These are called *stream standards* and refer to water sources like rivers, lakes, estuaries, oceans or ground water.

DRINKING WATER STANDARDS

In view of the direct consumption of water by human beings, the domestic water supply is considered to be most critical use of water. In India the agencies like Indian Council of Medical Research (ICMR), Bureau of Indian Standards (BIS) and Ministry of Works and Housing (MWH) have formulated certain drinking water standards which are being followed by different authorities. The World Health Organisation (WHO) has also laid down drinking water standards which are considered international standards (Table 131).

Table 131. Drinking water standards

Characteristics	World Health Organization(WHO)		Ministry of Works and Housing, (1975)	
	Highest desirable	Maximum permissible	Acceptable	Cause of rejection
1	2	3	4	5
PHYSICO-CHEMICAL				
Turbidity (J.T.U.)	5.0	25.0	2.5	10.0
Colour (Pt-scale)	5.0	50.0	5.0	25.0
Taste and Odour	nothing	disagr-eeable	unobjec-tionable	unobjec-tionable
pH	7.0-8.5	6.5-9.2	7.0-8.5	6.5-9.2
Total solids	500	1500	500	1500
Total hardness	100	500	200	600

1	2	3	4	5
Chlorides	200	600	200	1000
Sulphates (as SO_4)	200	400	200	400
Fluorides (as F)	1.0	1.5	1.0	1.5
Nitrates (as NO_3)	45	45	45	45
Calcium (as Ca)	75	200	75	200
Magnesium	·30	150	30	150
Iron (as Fe)	0.1	1.0	0.1	1.0
Manganese (as Mn)	0.05	0.5	0.05	0.5
Copper	0.05	1.0	0.05	1.5
Zinc	5.0	15.0	5.0	15.0
Phenolic compounds	0.001	0.002	0.001	0.002
Detergents, anionic	0.2	1.0	0.2	1.0
Mineral oil	0.01	0.30	0.01	0.30
Arsenic	0.05	0.05	0.05	0.05
Chromium (as Cr^{+6})	–	0.01	0.05	0.05
Cyanide	–	0.05	0.05	0.05
Lead	–	0.10	0.10	0.10
Selenium	–	0.01	0.01	0.01
Cadmium	–	0.01	0.01	0.01
Mercury	–	0.001	0.001	0.001
PCBs (μg/L)	–	0.2	0.2	0.2
Gross alfa-activity (PCi/L)	–	3.0	3.0	3.0
Gross beta-activity (PCi/L)	–	30.0	30.0	30.0

BACTERIOLOGICAL STANDARDS

W.H.O.	Ministry of Works and Housing
1	2
(a) *Water Entering Distribution System :* If disinfected, coliform count in any sample of 100 ml should be zero.	(a) Caliform count in any sample of 100 ml should be zero.
(b) *Water in the distribution system :* Ideally all samples	(b) Water in the distribution system shall satisfy all the three criteria indicated below:

1	2
taken from the distribution system including consumer's permises should be free from coliform organisms. Since in practice it is not always possible, following standards can be followed.	
(i) Throughout any year, 95% of the samples examined should not have any coliform organisms.	(i) E.coli count in 100 ml of any sample should be zero.
(ii) E.coli count in 100 ml of any samples should be zero.	(ii) Coliform organisms not more than 10 per 100 ml shall be present in any sample.
(iii) Coliform organisms not more than 10 per 100 ml shall be present in any sample.	(iii) Coliform organisms should not be detectable in 100 ml of any two consecutive samples of more than 50% of the samples collected for the year.
(iv) Coliform organisms should not be detectable in 100 ml of any two consecutive samples.	

Note : All the values are in mg/L except pH, otherwise stated.

IRRIGATION WATER STANDARDS

With regard to the quality of water for irrigation, the major parameters of concern are salinity denoted by dissolved solids and conductivity, potentially toxic trace elements, and herbicides. Besides the presence of sodium is also an important parameter, the excess quantities of which can deteriorate the soils. High volume of sodium may also damage the sensitive crops because of sodium phytotoxicity. The sodium in water can be denoted by percent sodium and *sodium absorption ratio* (SAR). These two parameters can be calculated by the following formulae. The values of individual parameters are taken in mEq/L.

$$\%\text{Sodium} = \frac{\text{Na}}{\text{Ca} + \text{Mg} + \text{K} + \text{Na}} \text{x} 100$$

$$\text{SAR} = \frac{\text{Na}}{\sqrt{\frac{\text{Ca} + \text{Mg}}{2}}}$$

Observations (Table 132) indicate the suitability of waters with different constituents, for irrigation. Tables 133 and 134 provide the limits for SAR and trace elements in irrigation water used continuously fro crops.

Table 132. Suitability of water with different constituents, for irrigation.

Class of water	TDS ppm	Sulphates ppm	Chlorides ppm	% Sodium	Boron ppm	E.C. μmho	Suitability for irrigation
I	0-700	0-192	0-142	0-60	0-0.5	0-750	Excellent to good for irrigation
II	700-2000	192-430	142-355	60-75	0.5-2.0	750-2250	Good to injurious, suitable only with permeable soil and moderate leaching, harmful to sensitive crops
III	>2000	>480	>335	>75	>2.0	>2250	Unfit for irrigation

Table 133. Suitability of water for irrigation with different values of SAR.

SAR	Suitability for irrigation
0-10	Suitable for all types of crops and all types of soils except for those crops which are highly sensitive to sodium.
10-18	Suitable for coarse textured or organic soil with good permeability. Relatively unsuitable in fine textured soil.
18-26	Harmful for almost all types of soils. Require good drainage, high leacing and gypsum addition.
> 26	Unsuitable for irrigation

Table 134. Trace elements limits for irrigation water in mg/L, used continuously for crops.

Aluminium	1. 0
Arsenic	1. 0
Boron	0. 75
Cadmium	0. 005
Chromium	5. 0
Cobalt	0. 2
Copper	0. 2
Lead	5. 0
Manganese	2. 0
Molybdenum	0. 005
Nickel	0. 5
Selenium	0. 05
Zinc	5. 0

STREAM WATER STANDARDS

Water quality objectives for freshwater take into account several major uses to which water is put like irrigation, drinking, industry, power generation, recreation and even for discharging waste water together with the fact that all water bodies or stretches are not necessarily required to meet all potential users. This has led to the concept of classification and zoning of water bodies which indicate that their quality has to meet the requirement of one or more of the above potential uses.

For each typical use, water quality critera are established that take into account the special constraints on water quality imposed by that use. Based on this, any water body or its stretch can be designated for some particular best use which can be termed as *designated best use.*

The water sources can be classified depending upon the designated best use of the water. The Central Pollution Control Board (CPCB) along with the State Pollution Control Boards has adopted a scheme of classification and zoning of water bodies

(Table 135). The water quality criteria for this classification has been shown in Table 136 for. freshwaters (Tyagi et al, 1991).

Table 135. Classification and zoning of water bodies (CPCB 1979)

Designated best use	Nomenclature for the class of water
FRESH WATERS	
1. Drinking water source without conventional treatment but after disinfection.	Class A
2. Out door bathing	Class B
3. Drinking water source with conventional treatment followed by disinfection	Class C
4. Propagation of wildlife, fisheries	Class D
5. Irrigation, industrial cooling, and controlled waste disposal.	Class E
SEA WATERS (Including Estuaries and Coastal waters)	
1. Salt pans, shell fishing, contact sport.	SWI
2. Commercial fishing, non-contact recreation.	SWII
3. Industrial cooling	SW III
4. Harbour	SW IV
5. Navigation, controlled waste disposal	SW V

Table 136. Water quality criteria for freshwater classification (CPCB, 1979)

Classes	Criteria
1	2
Class A	Dissolved oxygen (minimum 6 mg/L), BOD (maximum 2 mg/L), MPN of coliforms per 100 ml (maximum 50), pH (6. 5-8. 5)

1	2
Class B	Dissolved oxygen (minimum 5 mg/L), BOD (maximum 3 mg/L), MPN of coliform per 100 ml (maximum 500) pH (6. 5-8. 5)
Class C	Dissolved oxygen (minimum 4 mg/L), BOD (maximum 3 mg/L), MPN of coliforms per 100 ml (maximum 5000) pH (6. 0-9. 0)
Class D	Dissolved oxygen (minimum 4 mg/L) pH (6. 5-8. 5), Free ammonia as N (Maximum 1. 2 mg/L)
Class E	pH (6. 0-8. 5). Electrical conductivity (maximum µmhos 2, 250), Sodium absorption ratio. SAR (maximum 26), Boron (maximum 2 mg/L)

* Nomenclature of classes according to Table 132. If the coliforms are found to be more than the prescribed tolerance limits, the criteria for coliforms shall be satisfied if not more than 20% of the samples exceed the tolerance limits specified, and not more than 50% samples have values more than 4 times the tolerance limits. There should be no visible discharge of domestic and industrial wastes into class A waters. In case of Classes B and C the discharges shall be regulated/treated as to ensure the maintenance of the stream standards.

EFFLUENT STANDARDS

The effluent standards pertain to the quality of waste waters originating from community, agricultural operations, and industry. In general, these standards restrict the quality of pollutants in the effluents that set the desired degree of treatment. Some important effluent standards have been shown in Table 137.

Table 137. General standards for discharge of effluents (CPCB, 1995)

Sl. No.	Parameter	Standards			
		Inland surface water	Public sewers	Land for irrigation	Marine coastal area
1	2	3	4	5	6
1.	Colour and odour	*	*	*	*
2.	Suspended solids. mg/L. Max	100	600	200	(a) For process waste waster-100 (b) For cooling water effluent-10 per cent above total suspended matter of influent cooling water.
3.	Particle size of suspended solids,	Shall pass 850 micron IS Sieve			(a) Floatable solids. Max 3 mm (b) Settleable solids Max 850 microns.
4.	Dissolved solids (inorganic), mg/L Max.	2100	2100	2100	-
5.	pH value	5.5 to 9.0	5.5 to 9.0	5.5 to 9.0	5.5 to 9.0
6.	Temperature °C, Max.	Shall not exceed 40 in any section of the stream within 15 metres down stream from the effluent outlet	45 at the point of discharge	-	45 at the point of discharge
7.	Oil and grease, mg/L Max.	10	20	10	20
8.	Total residual chlorine, mg/L. Max.	1.0	-	-	1.0

1	2	3	4	5	6
9.	Ammonical nitrogen (as N). mg/L Max.	50	50	–	50
10.	Total Kjeldahl nitrogen (as N). mg/L Max.	100		–	100
11.	Free Ammonia (as NH_3), mg/L Max.	5.0	–	–	5. 0
12.	Biochemical oxygen demand (5 days at 20°C) Max.	30	350	100	100
13.	Chemical Oxygen demand, mg/L Max.	250	–	–	250
14.	Arsenic (as As) mg/L. Max.	0.2	0.2	0.2	0.2
15.	Mercury (As Hg) mg/L. Max.	001	0.01	–	0.01
16.	Lead (as Pb), mg/L Max.	0.1	1.0	–	1.0
17.	Cadmium (as Cd), mg/L Max.	2.0	1.0	–	2.0
18.	Hexavalent chromium (as Cr^{+6}) mg/L Max.	0.1	2.0	–	1.0
19.	Total chromium (as Cr), mg/L Max.	2.0	2.0	–	2.0
20.	Copper (as Cu), mg/L Max.	3.0	3.0	–	3.0
21.	Zinc (as Zn), mg/L Max.	5.0	15	–	15
22.	Selenium (as Se), mg/L Max.	0.05	0.05	–	0.05
23.	Nickel(csNi), mg/L Max.	3.0	3.0	–	5.0
24.	Boron (as B), mg/L Max.	2.0	2.0	2.0	–

1	2	3	4	5	6
25.	Percent sodium, Max.	–	60	60	–
26.	Residual sodium carbonate. mg/L Max.	–	–	5.0	–
27.	Cyanide (as CN), mg/LMax.	0.2	2.0	0.2	0.2
28.	Chloride (as Cl), mg/L Max.	1000	1000	600	–
29.	Fluoride (as F), mg/L Max.	2.0	15	–	15
30.	Dissolved Pbosphates (as P), mg/L Max.	5.0	–	–	–
31.	Sulphate (as SC_4), mg/L Max.	1000	1000	1000	–
32.	Sulphide (as S), mg/L Max.	2.0	–	–	5.0
33.	Pesticides	Absent	Absent	Absent	Absent
34.	Phenolic compounds (as C_6H_5OH) mg/L Max.	1.0	5.0	–	5.0
35.	Radioactive materials:				
	(a) Alpha emitters µC/mL, Max.	10^{-7}	10^{-7}	10^{-8}	10^{-7}
	(b) Beta emitters µC/mL Max.	10^{-6}	10^{-6}	10^{-7}	10^{-6}

* All efforts should be made to remove colour and unpleasant odour as far as practicable.

MINIMAL NATIONAL STANDARDS (MINAS)

The MINAS are industry specific effluent standards which are being evolved at the national level. Hence, State authorities are not required to relax on them except when the ambient water quality critera warrants their alteration to suit the location. This envisages the treatment of all waste waters to certain minimum standards regardless of the type of waste waters and their location. The minimum treatment that can be provided to any waste water aims at the removal of the pathogens, toxic substances, colloidal and dissolved organic solids, mineral oils and adjustment of pH.

The MINAS are evolved for different types of industries considering the treatability of waste waters by various unit processes available, cost involved for each stage in each process, performance of each stage of treatment in the process expressed as percentage removal of pollutants, and cost percentage for each process of the annual turnover of the industry. The formulation of MINAS is, therefore, linked to the techno-economic acceptability of the suggested treatment and ratio of total annual cost of pollution control to the annual turnover of the industry. The stage of the treatment whose cost remains under the critical percentage of annual turnover can be accepted as the minimal stage of treatment and the quantities of pollutants in this treated effluents as MINAS. What percentage of annual pollution control cost to the annual turnover should be considered critical an supercritical has to be decided by the industry committee.

The industry can be classified as soft if the pollution control cost remains below the critical percentage of annual turnover. The industry is hard, if this percentage is above supercritical percentage, and medium hard, when it fall between critical and supercritical percentages. For example, the paper-and pulp industry can be classified as a soft industry because the annual cost of pollution control remains below the critical percentage of total turnover for all the unit processes available for treatment. The minimum reduction of BOD and suspended solids possible by the best process is upto the limits of 30-50 mg/l and for COD upto 1000 mg/l, hence they can be considered MINAS for this industry.

Chapter-11

BIOTECHNOLOGY FOR WATER POLLUTION CONTROL

The undesirable waste characteristics of polluted water include the following: (i) Suspended solid and soluble organic compounds, which undergo progressive decomposition and thus result in oxygen depletion and production of noxious gases; (ii) Heavy metals, cyanides and other toxic organics which are deleterious to aquatic life; (iii) Undesirable levels of nitrogen and phosphorus, which enhances eutrophication (excessive plant growth, which kills animals due to deprivation of oxygen) and stimulate undesirable algal growth; (iv) Non-biodegradable chemical and volatile materials like Hydrogen sulphide and Sulphur dioxide.

The objective of biological treatment of waste water are to coagulate and remove the non-settleble colloidal solids and to stabilize the organic matter. They are considered better because they not only remove colour, but help in removal of carbonaceous organic matter in waste water which is measured as BOD (Biological Oxygen Demand), or COD (Chemical Oxygen Demand), or TOC (Total Organic Carbon), nitrification, denitrification and stabilization are the purposes of biological treatment.

Biological treatment can be carried out if the effluents are rich in unstable organic matter. The microbes break-up these unstable organic pollutants into stable products like carbon dioxide, carbon monoxide, ammonia, methane, hydrogen sulphide etc.

Industrial effluents vary in load, concentration of pollutants, toxic materials and are often nutritionally unbalanced. They may contain biodegradable or non-biodegradable or both types of pollutants. They show seasonal variations related with production.

Biological treatment of waste waters is a relatively neglected area and is also less properly understood and controlled. Many people do not give importance to choosing of right biological treatment method. Many people handle biological treatment in the most crude manner. Ignorance about the composition of effluents just cannot be excused.

Effluent treatment systems broadly fall into two categories: (i) Aerobic, and (ii) Anaerobic (Table 138).

Table 138. Major points of comparison between aerobic and anaerobic treatment of industrial effluents (Jogdand, 1995).

Aerobic Treatment	Anaerobic Treatment
1. Range of waste-waters can be treated.	1. Limitation in treatment applications.
2. Process stability and control.	2. Less process stability and control.
3. Require more, power input.	3. Require less power input.
4. Produce more sludge.	4. Produce less sludge.
5. Percentage BOD removed is more.	5. Percentage BOD removed is less.
6. Nitrogen removal is better.	6. Nitrogen removal poor.
7. Phosphorus removal is better.	7. Phosphorus removal poor.
8. Can cope up with low substrate level wastes.	8. Advantageous for high substrate level wastes.
9. Works in meophilic range of temperature.	9. Advantageous at higher temperatures.
10. —	10. Reductive dechlorination occurs.
11. Nitrification, denitrification phosphorus accumulation, ligninase activity occur.	11. —

AEROBIC BIOLOGICAL TREATMENT

They are based on the system of sewage treatment and are meant to handle easily biodegradable organic matter. when applying to industrial effluents, a careful treatability study should be done to determine the design and working parameters. The basic reaction in aerobic treatment plant is:

$$\text{Organic material} + O_2 \xrightarrow[\text{Other Nutrients}]{\text{Cells}} CO_2 + H_2O + \text{New Cells}$$

Microbial cells undergo progressive autooxidation of the cell mass:

$$\text{Cells} + O_2 \longrightarrow CO_2 + H_2O + NH_3$$

Lagoons and low rate biological filters have limited industrial applications but activated sludge and fixed film systems of different types are widely used for industrial effluents. Advanced activated sludge systems use pure oxygen instead of air and can operate at higher biomass concentration.

ACTIVATED SLUDGE PROCESS

It operates as a homogeneous continuous culture and is aerated. Biosorption and floculation removes the organic matter rapidly while oxidation and biosynthesis proceed at a lower rate. Subsequently flocs settle into the next stage of secondary sedimentation tank. A portion of this floc may be returned as inoculum. BOD and suspended solids are reduced by 85-95 percent. Organisms in activated sludge are similar to that in percolating filters. In the activated sludge, conditions are not suitable for macro-invertibrates grazer population. Resultantly, activated sludge plants do not suffer from nuisance by flies. Also fungi are less domonant. So less sludge bulking. Protozoans are abundant in activated sludge. Nematodes and rotifers are small in number.

The contents of the reaction vessel are referred to as Mixed Liquor Suspended Solids (MLSS) or Mixed Liquor Volatile Suspended Solids (MLVSS) and consists mostly of micro-organisms and inert and non-biodegradable suspended matter.

The original activated sludge process was introduced in 1914. It has certain drawbacks: (1) High running costs; (2) Difficult to operate and maintain; (3) Produce large surplus biomass. There are various modifications of activated sludge process:

(A) Tapered aeration

Here aeration capacity is related to demand and it is less at the outlet than at the inlet.

(B) Step aeration

Here feeding as well as aeration is done at steps in the system throughout length of the tank (Figure 47).

(C) Contact stabilization

Here returned sludge is aerated to encourage organisms to utilize any stored nutrient More wastes digested. Sludge volume is reduced through aerobic digester stage. It is similar in principle to the extended aeration treatment. For extended aeration treatment, aeration and mixing of sludge and effluent is done in the same unit which is conducted separate in contact stabilization.

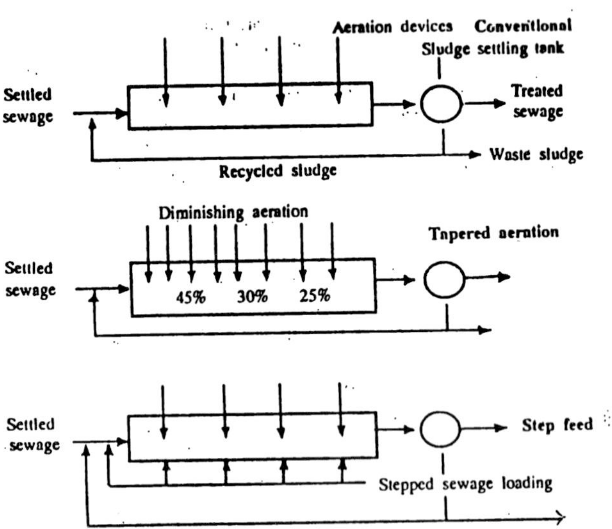

Figure 47. Modifications of activated sludge process.

(D) Advanced activated sludge process

Most of them operate with pure oxygen. So they can operate at a higher biomass concentration. This reduces residence time and bulking (i.e. excessive growth of filamentous bacteria and fungi which may inhibit sludge setting) is inhibited.

Powdered activated carbon (PAG) may be added as additive in the activated sludge process. This absorbs organic compounds which cannot be degraded. It also reduces colour and metal in the effluent. It reduces effective level of inhibitors. PAC can also overcome toxicity to nitrification in activated sludge plant treating coking plant effluents.

Activated sludge is still considered as a cost-effective technology. The common problems of activated sludge are nutritional deficiency, pH out of 6 - 8.5, temperature beyond mesophilic range, poor flocculation of biomass due to perhaps very few different degradable substrates and few species of bacteria.

Carrier Activated Sludge Process

Activated sludge systems modified by the addition of inert suspended solid particles (carriers) are widely used. These CASP systems combine the characteristics of fluidized bed systems and conventional activated sludge systems. Here biomass concentration may vary from 8000 to 30,000 mg/l as MLVSS. Disadvantages experienced in fludized bed systems like ineffective suspended solid removal and operational difficulties are overcome here. CASP requires minor modifications for existing process installations conversion. Biomass is attached as well as in flocs form. Surface area is considerably increased. Biomass concentration is much more than conventional activated sludge process. Carriers may include sand, plastics, glass, powdered activated carbon, clay etc. Microcarriers if used help flocs formation in addition to providing of attached growth support. Biological flocs capture particulate matter and hence effectively remove fine suspended solids.

ADVANCED ACTIVATED SLUDGE PROCESS

(High Purity Oxygen Systems)

Closed tank systems	Open tank system
(1) Monitoring and Maintenance is difficult.	(1) Ease of access for monitoring and maintenance.
(2) Less treatment capacity.	(2) Treatment capacity is increased.
(3) CO_2 accumulates in Head space. Hazardous.	(3) Allows escapes of CO_2 from mixed liquor to atmosphere.
(4) Examples — (i) UNOX system (ii) OASES system (iii) Forced Free-full 'F^{3}O' system (iv) Marox system	(4) Examples – (i) Vitox system (ii) Megox system (iii) Primox system (iv) Simplox system
(5) It prevents loss of unutilised oxygen and permits recirculation	(5) Unutilised oxygen is lost.
(6) It prevents odour and spray nuisance which are normally associated with activated sludge processes.	(6) Odour and spray nuisance exists.
(7) Combustible materials like oil, fat, grease burn vigorously in O_2 rich atmosphere. Hydrocarbons should not be allowed to accumulate in head space by doing purging and also oxygen should not come in contact with hydrocarbon lubricants. This will reduce possible hazards. Chances of hazards are also low as system is operated at ambient temperature and low pressure and gas phase being saturated with water vapour.	(7) No possibility of hazard as no accumulation of hydrocarbons in headscape.

BIOLOGICAL FILTERS - FIXED FILM SYSTEMS

Micro-organisms are attached to an inert supporting medium which is packed into a tower or tank. There are some varations like rotating biological contactor's which basically are aerobic systems with fixed film of micro-organisms growing on discs on a rotating shaft. Even distribution of effluent is done in the reactor and air is introduced from the bottom vents and passes it across the media bed as effluent percolates or distributes.

Microbial slime develops on media support by using organic matter and oxygen. When the thickness of the slime increases, extra biomass sloughs off. This sludge is collected by gravity in a sedimentation tank. Sludge is later on treated and disposed off the same way as for activated sludge.

The first trickling filter was put in operation in England in 1893. Filter media used in trickling filter normally consists of rocks varying in size from 25-100 mm in diameter. The depth of the rock varies with design and ranges from 3 to 8 ft. The use of plastic medium in trickiling filters is relatively a recent innovation and tanks are square-shaped with depths of 30-40 ft. Rock filter beds are circular with a rotating arm distributing liquid effluent at the top of the bed. Filters have an underdrain system to collect sludge and treated effluents. Organic matter is decomposed by micro-organisms grown as film on inert support. Development of slime, metabolic activities of micro-organisms in slime, increase in thickness of slime, detachment, of slime when micro-organisms near to medium reach endogenous phase (lose the ability to cling), formation of new slime is a continuous process occurring in a cyclic manner. Facultative bacteria like *Achromobacter, Flavobacterium, Pseudomonas, Adcaligenes,* filamentous forms like *Sphaerotilus natans, Baggiatoa* and at lower levels of bed nitrifying bacteria like *Nitrosomonas, Nitrobacter* are common. Also few fungi, algae and protozoa are present.

Percolating liquid washes the slime off the medium. Sloughing off depends on organic and hydraulic loading of the filter. Hydraulic loading accounts for shear velocities and organic loading accounts for the rate of metabolism in the slime layer. On the basis of hydraulic and organic loading rates, filters are divided into two classes: low rate and high rate.

ROTATING BIOLOGICAL CONTACTORS (RBC)

This is one of the principle types of fixed film-moving medium systems used as a digester. Biological growth is established on disc surface made of polystyrene, polyethylene, polypropylene, stainless steel, cement, aluminium, glass, PVC, rubber, teflon, wood, wire screens etc.

Discs are 2-3 meter in diameter, 10-20 mm wide and mounted in a horizontal shaft. The distance between adjacent discs is 20 mm. Discs are partly submerged (40% area in the medium). Discs are rotated at 1-7 revolutions per minute. Air is sparged to reduce the risk of anaerobiosis in multiple unit systems. Retention time of medium is comparable with percolating time. Biomass on discs is 200 gm dry weight per square metre of disc surface. Biological growth is 2.4 mm thick (Figure 48).

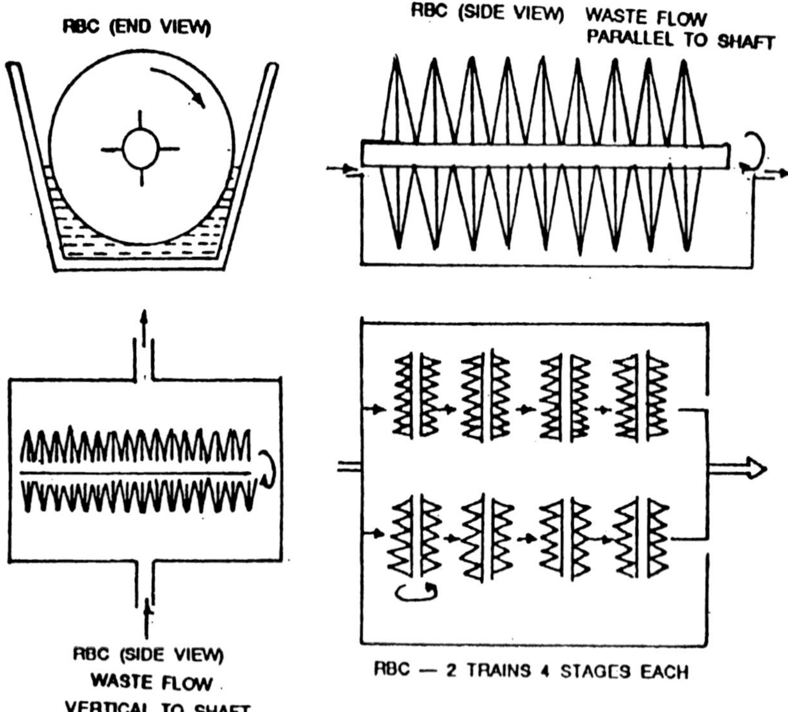

Figure 48. Rotating biological contactors (SBC).

ADVANTAGES OF RBC

(1) They are simple to operate, have a low maintenance, (only lubrication is required).

(2) It accommodates shock loadings.

(3) It does not have the channeling problem as in the percolators.

(4) It has reduced power costs (no sludge recycling) and less operational and maintenance costs, too.

(5) It requires less space than activated sludge plant.

(6) Fast startup, efficient mixing, litlle sloughing off of biomass.

(7) Effluent quality achieved is as good as after tertiary treatment.

(8) Foaming, aerosol, airstripping is reduced.

(9) Problems faced in trickling filters (percolators) like clogging, ponding, filler flies are eliminated.

(10) High waste-water temperature, oxygenation may be easier with RBC than with the activated sludge system.

(11) Lower head loss compared to the trickling filters.

(12) Process is self-regulating with respect to cell retention time and stability of process is a plus point as biomass neatly attaches to media support.

DISADVANTAGES/PROBLEMS WITH RBC

(1) Lack of operational control due to oversimplicity of the process.

(2) Process is still new and people have less experience.

(3) There are difficulties in kinetic evaluation due to the complexity of interactions between biomass and solid, liquid, gaseous phase.

(4) Undesirable heavy growth of *Baggiatoa, Thiothrix* may occur. More sludge production.

(5) High total Vs. Soluble BOD in RBC effluents.

(6) Dissolved oxygen may remain limited.

(7) Odour problem.

(8) Enclosures required for RBC to protect it from sun, cold and heavy precipitation.

RBC systems find applications for effluent treatment for the following purposes: (a) secondary treatment, (b) nitrification, (c) landfill leachate and runoff, (d) pretreatment of water supplies, (e) denitrification, and (f) phosphate removal etc.

FLUIDISED BED REACTORS (FBR)

It is combination of attached growth (percolating filter) and suspended growth (activated sludge) systems. Biological slime film is developed and maintained on a solid support medium consisting of particles small enough to be maintained in suspension by the upward flow of liquid being treated. Support medium particles neither sink nor outflow. Reactors are generally cylindrical with perforated distribution plates and tapered or conical entry sections. Fluidization of support particles is allowed but clumping prevented. (Figure 49).

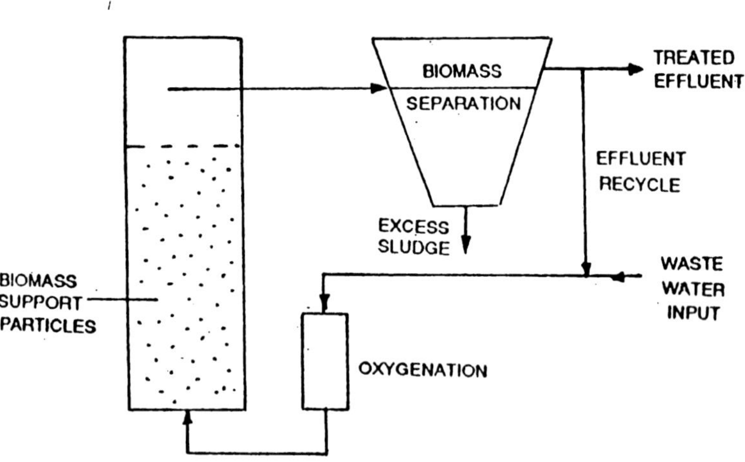

Figure 49. Fluidised bed reactor.

Support media used may be sand, carbon, flyash, anthracite, glass calcinated clay with particle size 0.2 to 0.3 mm. Factors important in the selection of support particles are particle density, particle size, size distribution, surface area, sphericity, composition of material, hardness, bulk density, cost. These mineral media support growth only on the outer surface. Smaller the size, larger is the surface of the film action but lower is the settling rate is reactor. There are also fabricated media developed which allow biomass to grow within the porous internal structure.

ADVANTAGES OF FBRS

(1) Compact small reactors; so space saved.

(2) Easy to install, easy for shipment. Only piping to be done on site.

(3) It is a quick, economical expansion alternative in treatment plant.

(4) Modules can be easily added to the existing system.

(5) It is suitable lor high-strength industrial waste-water with >1000 COD.

(6) Low capital and operating costs.

(7) Higher reactor biomnss concentration than other suspended growth or fixed film systems.

(8) Highly efficient as compared to other systems.

(9) Secondary clarifiers can be possibly eliminated for solid separation and recycling.

(10) High exposed liquid/biofilm surface area.

(11) Sludge transport and disposal costs are lower.

(12) Low biomass sludge production.

INVERSE FLUDISED BED BIOFILM REACTOR (IFBBR)

Fludised bed reactors that are in use today are all operating with upflow systems, consisting of gas-solid, liquid-solid or gas-liquid-solid phases.. Density of bioparticles is higher than the density of the medium. Upflow biofluidization is inconvenient for certain applications carrying aerobic processes. Collisions between the bioparticles and shear stress affects biofilm formation and then biofilm thickness. Inverse fluidization (Figure 50)

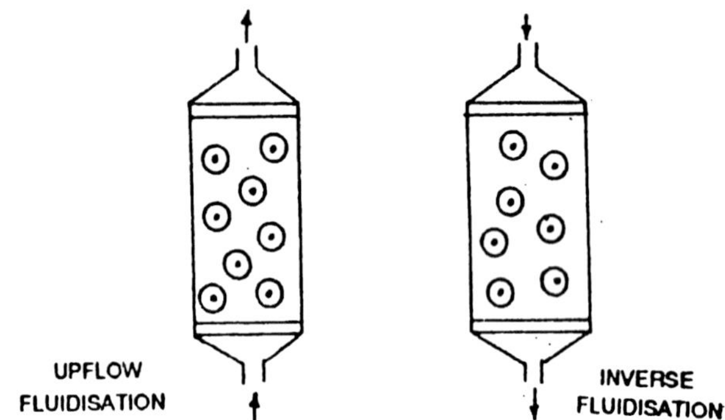

Figure 50. Upflow and inverse fluidisation bed biofilm reactor.

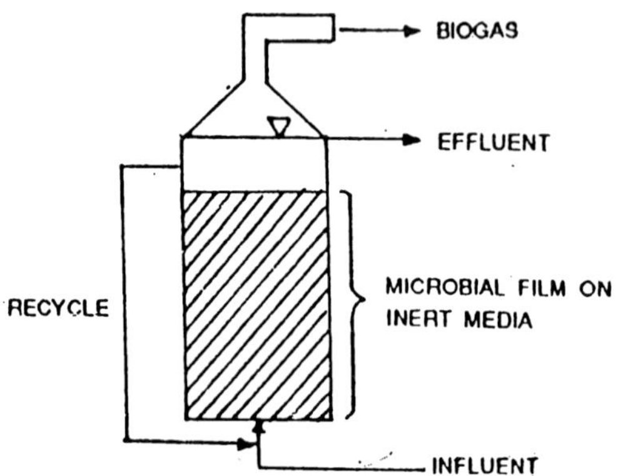

Figure 51. Expanded bed reactor.

removes the drawbacks of normal up flow FBR and gives a higher performance. Three advantages of inverse biofluidization are: (a) effective and simple control of biofilm thickness, (b) large specific support surface area, and (c) fast biofilm formation.

EXPANDED BED REACTOR (EBR)

The basic mode of operation is similar to the packed filters and fluidized bed processes. Fluidized processes were developed earlier and have diverse applications. The expanded bed system uses the operating mode of fluidized bed reactor. For biological applications, an expanded bed is a different process than fluidized bed in certain aspects. These are: (a) velocities to maintain the delicate attached living film, (b) separation, retardation and biocoagulation of fine suspended solids, (c) achievement of maximum biomass concentration (Figure 51).

ANAEROBIC BIOLOGICAL TREATMENT

Anaerobic digestion is microbial fermentation by which in the absence of high levels of nitrates and sulphates, organic matter is converted to carbon dioxide and methane.

Anaerobic digestion has long been used for sludge digestion and stabilization in sewage treatment process. Anaerobic digestion was seriously considered for the treatment of high strength industrial effluents initially. Earlier, anaerobic digestion was considered sensitive to toxicity and less reliable in performance in terms of COD reduction. This misconception has been proved untrue in recent years. Also increase in costs of energy and sludge disposal has deterred the use of aerobic biological treatment procedures which are energy consuming sludge producing and acting towards net loss of material. As against this anaerobic digestion of effluents has several advantages over aerobic digestion. These are:

(1) Very low sludge production.

(2) Lower consumption of energy.

(3) Production of methane which has high calorific value.

(4) Process ran operate at high organic loading rate.

(5) No environmental nuisance of odour, aerosol as is the case with aerobic treatments.

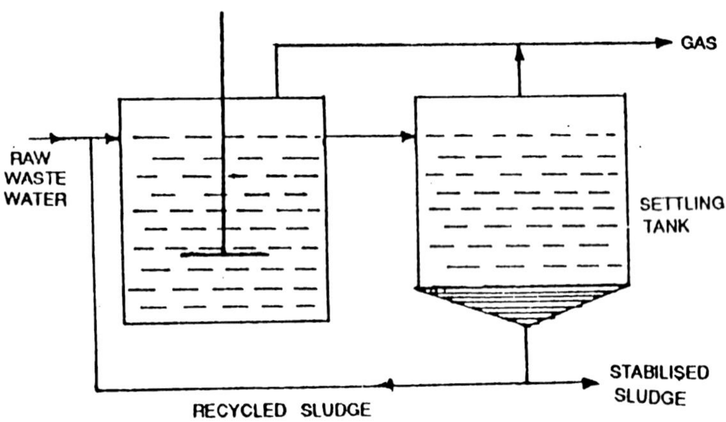

Figure 52. Anaerobic contact digester.

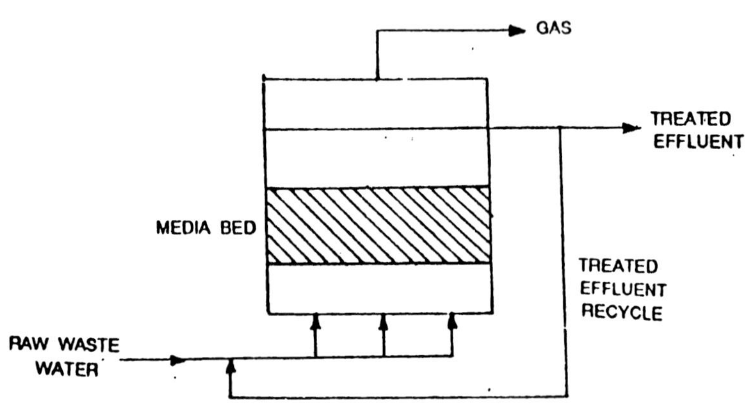

Figure 53. Anaerobic packed bed reactor.

(6) Micro-organisms can remain dormant for several months and become operational within a week of start up. This is suitable for seasonally produced waste-waters.

ANAEROBIC CONTACT DIGESTERS

This is an aerobic equvalent of activated sludge process. It consists of a stirred tank and a tank under anaerobic conditions. The output of completely stirred tank (digester) is settled under anaerobic conditions and a part of settled sludge in returned to the digester. This results in the concentration of the sludge and longer retention time. This enables retaining of methanogenic organisms over a wide range of loading. Separation of bacteria is hindered by gassing of the effluent, hence the effluent is usually degassed before settling of the biomass.

Anaerobic contact digesters (Figure 52) are currently used for treating effluents from sugar processing, distilleries, citric acid and yeast production, industries producing saurkraut, canned vegetables, pectin, starch and meat products, farm slurries. Contact digesters are not much affected by the suspended solids in the feed as it happens in retained biomass types of digesters.

One of the main problems with contact digesters is the poor settlement of solids because of attachment of product gas to solid particles.

PACKED BED REACTORS (PACKED COLUMN REACTORS)

They are simple in design, easy to construct and operate (Figure 53). Organisms are contained within the packing medium in an enclosed vessel and liquid wastes pass upwards. Organisms do not form slime on packing. Regular backwashing will prevent clogging. This will also prevent high concentration of suspended solids shoughing continuously into the effluent. The treatment (removal) rate is directly related to surface area of packing.

Polystyrene spheres in anaerobic filter or Neptune-Microfloc filter containing sand, silica, anthracite coal are used.

ANAEROBIC BAFFLED DIGESTERS

These reactors have walls across the tank built from the top to the bottom, so that effluent flowing along the tank has to

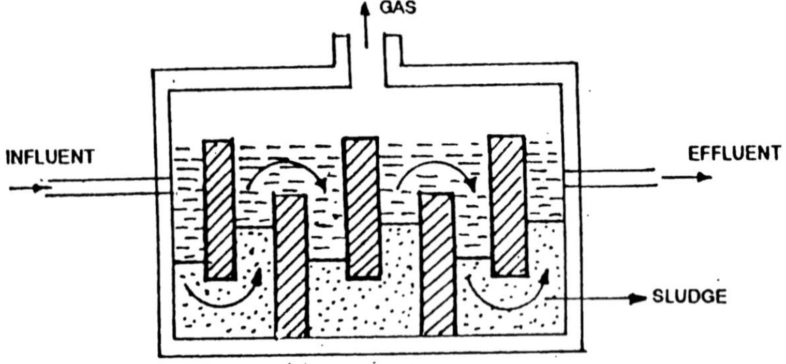

Figure 54. Anaerobic baffled digesters.

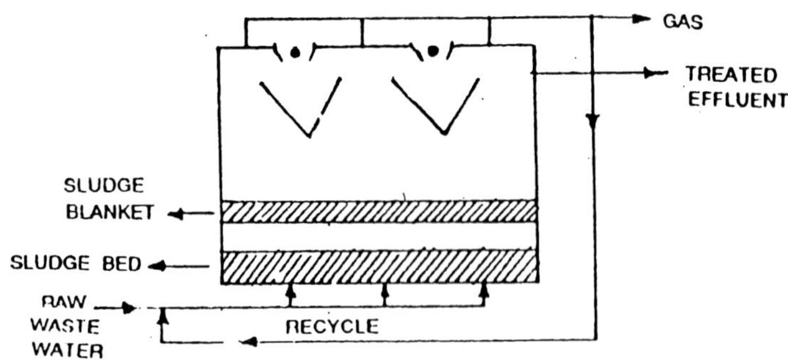

Figure 55. Upflow anaerobic sludge blanket reactor.

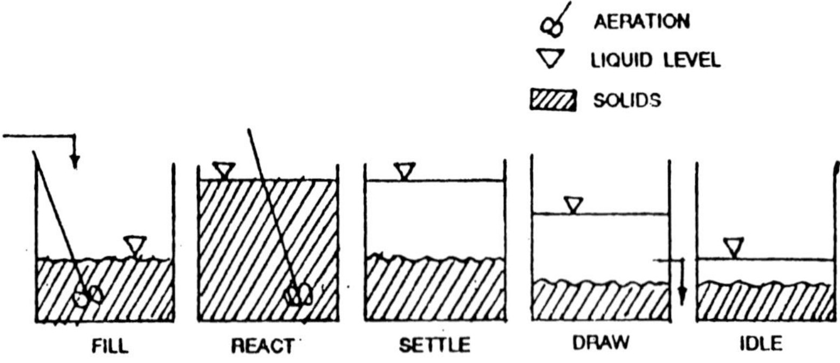

Figure 56. Periodic biological reactors.

go alternately under and over the baffles. The baffles tend to keep the bacteria in the tank and also help to prevent problems with the floating solids (Figure 54).

UPFLOW ANAEROBIC SLUDGE BLANKET REACTOR

These reactors require the active bacteria in the form of high-density granular sludge which is retained in the digester tank despite gassing and upflow velocity of effluent (Figure 55). Sludge granulation is complex and not yet fully understood. Initially 10-15 % inocula of granular sludge is required. Hydrodynamics of the digester created by feed distribution and the shape of digester are important in retaining the correct granular form. Well-adapted sludge may be sufficient in 1% volume only. The key element in the feed substrate for successful granule formation are calcium, phosphorus, magnesium, ammonia, aluminium, silicon. A large population of filamentous micro-organisms (e.g., *Mothanothrix* spp) is also essential. Granular sludge develops with the dissolved wastes. Baffles in the digester, promote gas solid separation along with the shape of digester and upward liquid velocity. Baffles provide the surface area which reduces the upwards flow velocity and promotes biomass flocculation. Baffles are corrosion-resistant reinforced plastic. Long chain fatty acids prove toxic to methanogenic bacteria and addition of calcium chloride can reduce this toxicity and can help better granulation.

PERIODIC BIOLOGICAL REACTORS

One of the most recent advances in biological treatment processes is the use of Periodic biological reactors (Figure 56). These reactors markedly reduce the problems associated with effluent varations and eliminate sludge recycling.

MEMBRANE BIOREACTORS

Membrane separation is useful to remove, the inhibitory effects of biodegradable pollutants. Membrane bioreactor for volatile organic compounds (Figure 57) consists effluent inlet chamber, from where the effluents pass through membrane and reach over biofilm. Degradation occurs in biofilm. Aerating gas and waste water do not come in direct contact. The use of specialised cultures in conjunction with bioreactors dedicated to treatment of point source discharges of particular waste water is advocated.

USE OF IMMOBILISED ENZYMES OR MICROBIAL CELLS FOR EFFLUENT TREATMENT

Suspended growth and fixed film systems are the two most common types of treatment used for effluents. Both are based on microbial growth, metabolic activities of micro-organisms, eventual death of organisms sludge production and bioconversions occurring during the whole process. Immobilisation of biocatalyst (Figure 58) cells or enzymes can add to the advantages of biological treatment system.

An immobilised biocatalyst can be defined as a biocatalyst for which movement in space in completely or severally restricted to form a distinct phase within the bulk phase in which substrate, effector, inhibitor molecules are dispersed and their exchange is possible.

WHY ENZYMES ARE SUITABLE AS A BIOCATALYST?

(1) Enzymes require low energy inputs.

(2) Enzymes have low heat of reaction.

(3) Enzymes are heat labile, so it is easy to terminate reaction.

(4) Enzymes require much less complex media.

(5) Less wastes are produced.

(6) Continuous operation is easier.

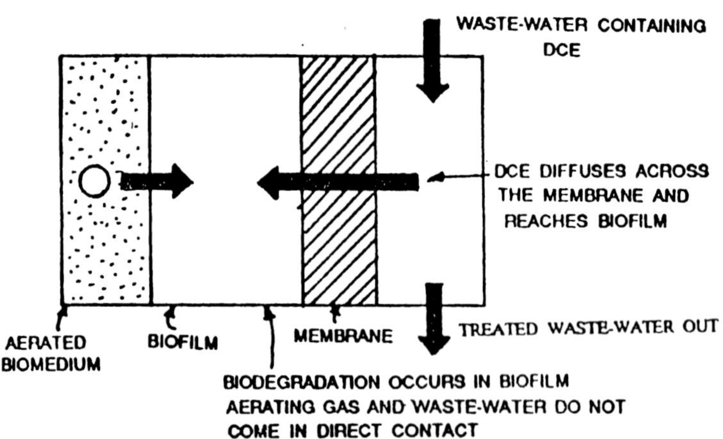

Figure 57. Membrane bioreactor for Volatile Organic Compounds.

WHY CELLS ARE SUITABLE AS A BIOCATALYST (SOURCE)?

(1) Enzymes when within the cells are better protected from denaturation.

(2) Degradation activity may be composed of a series of reactions working in concert and it will be difficult to develop different systems for different enzymes working together.

(3) Isolating, purifying the enzymes and then using them is expensive. Also chances of denaturation are more.

(4) Some enzymes occur as induced enzymes. Exploitation of such enzymes is easier if cells are used.

(5) Degradation rates and resistance to toxic pollutants will be great when cells as monoculture or mixed culture are used rather than using isolated enzymes.

ADVANTAGES OF IMMOBILISATION OF A BIOCATALYST

(1) Reuse or continuous use of cells enhances overall efficiency.

(2) Biocatalyst (cells/enzymes) docs not contaminate the product.

(3) Cells are more evenly dispersed by immobilisation, so diffusional restrictions are minimised.

(4) Concentration of biocatalyst possibly is more, so smaller size reactors can be used.

(5) Immobilised cells are used with more ease to exploit the kinetic features of continuously-stirred and packed bed reactors.

METHODS OF IMMOBILISATION OF A BIOCATALYST

(1) Entrapment in polymer matrix.

(2) Adsorption of charged biocatalyst on oppositely-charged support material.

(3) Covalent attachment to chemically-activated supports.

(4) Encapsulation inside semipermeable membrane.

(5) Aggregation of biocatalyst into floes.

(6) Biospecific attachment to supports by means of lectins etc.

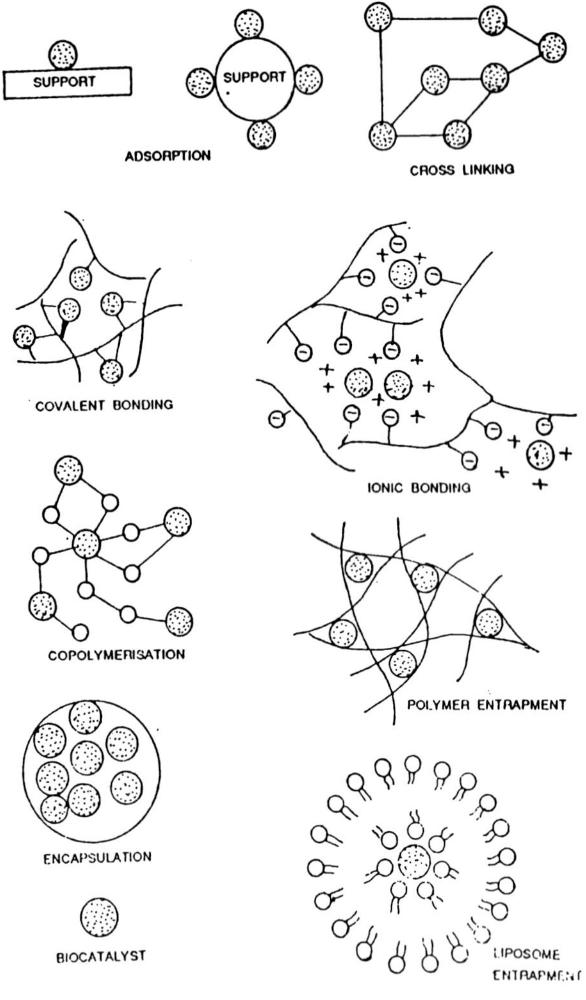

Figure 58. Methods of immobilisation of biocatalyst.

It will be out of place to discuss the details of various methods of immobilisation and their advantages and disadvantages. But the general advantages of immobilisation are already mentioned and methods of immobilisation are illustrated in Fig. 58.

APPLICATIONS OF IMMOBILISED CELLS AND ENZYEMS IN WASTE-WATER TREATMENT FALLS INTO FOUR CATEGORIES:

(a) BOD/COD reduction.

(b) Specific pollutant detoxification.

(c) As biosensors.

(d) For bioconvcrsion of waste to get specific products.

Flocculation of cells that occurs in activated sludge process or cells that are trapped on a slime layer in trickling filler beds may he considered as crude or preliminary way of applying/ using immobilized bio-catalyst. Improvements in these immoblisation techniques and purposeful procedures of immobilisation have proved useful in waste-water treatment. Natural flocculation of cells in suspended growth systems am he improved by the addition of synthetic polyelectrolytes.

Chapter-12

WATER QUALITY MANAGEMENT

The term water quality management has come into vogue in recent years as part of the current environmental crises. Increasing water quality problems and increasing demands for the beneficial use of water have combined to make it impossible that water quality be truely managed on a scale never before considered. In such instances, it is not unreasonable to say that the traditional approach to water quality management are no longer very satisfactory. In case of developing countries in particular, where financial supports are severely limited, it is necessary to analyse and advise the strategies of water quality management which would attain a reasonably acceptable water quality, at a minimum economic burden.

Water quality management is seldom defined in any regorous way, before proceeding with the analysis, it is advisable to establish the boundaries of what is meant by water quality management. Water quality management can be defined as an organised effort to maintain or improve the ambient water quality of the aquatic bodies, such that their most beneficiary use is not adversely affected, by exercising a control over the wastewater loads and the aquatic bodies themselves. The control over the riverine systems could include flow regulation, instream aeration etc. and control over the wastewater loads could consist of treatment for removal of pollutants such as organics, inorganics, heat etc. at strategic locations and with or without by-pass combinations. The system of control could be through the imposition of standards or charges, but imposition of standards alone is considered as a control system and a review of charges approach is excluded.

The organisation may take many different forms. Two possible extremes are management by a single authority to control a number of separate wastewater discharges and management by a number of authorities to control a single wastewater discharge - which is obtained by accumulating all the wastewater discharges entering the river basin. The concept of regionalisation is essentially related to the nature of such organisations and looks for a plausible option between these two extremes. Regionalisation has several promising features and these have been widely discussed notably by Kneese (1964), Yao (1973), Ludwig and Storrs (1973), and Okun (1977). Regionalisation brings forth a lot of possible alternatives and a need therefore arises to develop methodologies to choose the most optimum one. It is obvious that the techniques of mathematical programming and optimization are of great help in such a situation as it is often impossible to select the optimum solution based solely on the engineering judgements. Needless to state, therefore, optimization has become rather a rather a necessary step in regional water quality management.

In 1960-1980, almost a herculean effort has been made by a number of research workers to develop optimization models in regional water quality management. Although only a few have considered all the possible alternatives at a time, good amount of progress has been made where important alternatives such as wastewater treatment, wastewater transport at strategic locations, flow augmentation, in-stream aeration etc. have been considered. Models have also been proposed in the stochastic environment to come closer to the real world situations.

THE DEGREE OF TREATMENT PROBLEM

The degree of treatment problem can be described as minimizing the sum of the pollutant removal costs (in most instances BOD), for each wastewater discharge, subject to the constraints that the pollutant concentration, due to the combined efforts of the upstream discharges, does not drop below a preset standard for a mesh of constraint points in each reach.

A number of variations of degree of treatment problem have been presented such as uniform treatment, where the degree of treatment was same within the preset zones but the zones could have different degrees of treatment; least cost solution, where the degree of treatment could be different for each discharger and in this case the problem formulation would be the same as described for the general degree of treatment problem.

THE LOCATION PROBLEM

The degree of treatment is not the only potential alternative while searching for the optimum solution in regional water quality management. It is just one of the most important alternatives and a due consideration to the location of the dischaege should also be given while formulating the optimization problem.

This fact was soon realised in 1970. The consideration of locations introduced a complexity in the problem formulation as it was possible to have combined wastewater treatment units at strategic locations thereby taking advantage of the economies of scale at the cost of added transport facilities. Keegan and Leed (1970) considered both location as well as the degree of treatment in the optimization model. Graves et al (1972) presented an improvement, which considered both degree of treatment and location problems for the Delaware estuary.

CONSIDERATION OF THE RIVER CONTROL

River control in terms of flow augmentation and instream aeration is one of the important options in regional water quality management. A model considering the degree of treatment, location, instream aeration and flow augmentation all at a time was presented by McNamara (1976). However, little useful work has been done in this area and demonstrations of these models to the real world situations are necessary in order to arrive at some meaningful conclusions.

RELIABILITY OPTIMIZATION

The management models discussed earlier were all steady-state models, where all the parameters in the system such as wastewater flow, BOD, stream flow etc took deterministic values, that is, they did not change over time. In the real world, however, these parameters vary continuously with time and are thus stochastic in nature. The deterministic models are often adequate to evaluate the overall long range planning policies but their application becomes severely restricted when dynamic and short range policies are to be evaluated. Stochastic models are necessary in a large class of water quality management problems especially the ones which need the optimization of the operating strategies.

Once the problem setting is carried out in the stochastic environment, it is understood that the behaviour of the system is not upto the expectations all the time. In water quality management this would mean that for a certain fraction of time, the water quality may not be satisfactory. If reliability is defined as a measure of the effectiveness in terms of the attainment of the water quality objectives, then the problem of stochastic analysis can be related to the assessment of reliability. An increase in the reliability corresponds to an increase in the costs or investments. Reliability thus directly confronts with the economics of the system, and thus forms an important criterion for the optimum then mere cost minimization.

Shih (1976) presented first regional water quality management model which explicitly considered reliability optimization. He considered the stream flow, wastewater flow, effluent BOD and stream BOD as stochastic parameters at a time. The costs included both the capital as well as operations, maintenance and replacement costs applicable for the waste-water treatment units.

Reliability optimization is, however, in its infancy and a considerable amount of research is needed to draw some more meaningful conclusions. The role of optimization continues to remain as a cursory aid to the planning mechanism and managerial decisions in the area of regional water quality management.

WASTE WATER MANAGEMENT

It has been observed that agricultural, industrial and domestic use of water and disposal of untreated effluents, deteriorates the surface and underground water quality. Physico-chemical characteristics of selected industrial effluents and sewage in India (Table 139) show that the pH was highly alkaline

Table 139. Physico-chemical characteristics of selected industrial effluents and sewage of India (NEERI, 1975)

Industry	pH	Suspended solids (SS) (mg/l)	BOD (20°C) (mg/l)	COD (mg/l)	Miscellaneous constituents	Pollutional aspects
1. Sugar Factory	7.8-8.0	1500-1832	650-820	60-98		Large volume high pH, SS, colour and toxicity.
2. Straw board	7.5-12.9	3000	2000	5000	—	High pH SS and BOD
3. Tannery	9.5	3200	7000	—	Chromium 15-20mg/l	High BOD SS Cr and colour
4. Collon textiles	8.0-11.0	30-50	200-600	—	Detergents dyes chromium[3] mg/1	Alkali BOD dyes and varying chemical quality
5. Distillery	4.3	4000	29000	65000	Oil and grease	High BOD SS grease and ready putrescibility and low pH
6. Steel mill coke ovens	—	—	630	—	Grease oil and tar NH_3/N 1000 mg/l phenols 1300 mg/1 ONS	Highly toxic due to phenols cyanides and ammonia
7. Steel (finished)	—	310	280	—	Phenol 97.5 mg/1, NH_3/N 440 mg/1	— do —
8. Refinery	—	—	200	—	180 mg/1 and sulphidcs oil 30 mg/1 phenol 30 mg/1.	Mineral oil and phenols
9. Fertilizer Ammonia and Urea	8.0	3700	30	330	NH_3/N 510 mg/1 arsenic	Toxic due to free ammonia and promotes eutrophication
10. Dairy	8.0	690	816	1340	Oil and grease	High BOD SS grease and ready putrescibility
11. Sewage	7-8	200	350	500	Ammonia Albuminoid	

(7.5 - 12.9) in straw board effluents and highly acidic (4.3) in distillary effluents. Suspended solids were maximum (4000 mg/l) in distillary effluents and minimum (30-50 mg/l) in cotton textile effluents. Biological oxygen demand and chemical oxygen demand were maximum (29000 and 65000 mg/1 respectively) in distillery effluents, while these values were minimum (30 and 330 mg/1 respectively) in ammonia and urea industries effluents. Besides the above mentioned parameters, some of the constituents such as chromium, detergents, oil and grease, phenols, arsenic etc. were also disposed of by some particular industries.

The jute and textile mills in West Bengal have been polluting the Hooghly river, a stretch of 100 km has been found to contain 350 outfalls on either side of the river pouring large quantities of industrial effluents (approximately 20 x 10^6 m^3/day).

Industrial waste and sewage combine to pollute the Ganga river at Kanpur. These include effluents from tanneries, textile, wood and jute mills along the river. Gomti river at Lucknow is contaminated over a stretch downstream Lucknow from paper and pulp mill waste water and raw sewage.

Frequent power breaks in Delhi brings seven sewage treatment units to a stand still, resulting in dense sewage going into the Yamuna river in place of treated one. The statement that "Agra may be drinking Delhi's sewage" is well documented by an analysis of this river water along 195 km downstream to Agra, which reveals increasing deterioration in its quality (Chauhan, 1977). The Najafgarh drain, came to be known as "Sorrow of Delhi" after the famous Jaundice epidemic of Delhi in 1955, is still a "Sorrow of Yamuna", as the river continues to receive 3,18,180 m^3 /day of sewage and industrial effluents from the drain (Chakrabarty and Gupta, 1980).

Kalu river in Maharashtra is contaminated by nearly 100 industries, chiefly chemical and rayon (Sharma, 1977). Similarly, Krishna river in Maharashtra was receiving large amount of industrial effluents every day. Boralkar et al (1981) reported that there was an increase in the salt build up at an alarming rate in this river, as a result of several hundred hectares of land have already become worthless.

The addition of industrial and urban effluents to the cauvery river has resulted in a gradual buildup of total alkalinity, hardness, chlorides etc. (Table 140), creating polluted and organically enriched zone. Cauvery river is the major source of drinking water in Tamil Nadu. Polluted water of this river has caused endemic waterborne diseases in many towns (Pachpurkar, 1976; Paramasivam and Sreenivasan, 1981).

Table 140. Mean values of some important physico-chemical parameters of Cauvery river system (Sampath et al. 1981)

Sampling point	Effluent outfall	Surface water temp °C	DO (ppm)	BOD (ppm)	Total alkalinity (ppm)	Chloride (ppm)	Total organic carbon (ppm)	Total hard-ness (ppm)
1. Komara palayam	Sewage	28.8	9.7	6.2	136	30.8	7.00	113
2. Erode	Textile	29.2	7.7	3.8	130	24.5	6.25	102
3. Palli palayam	Pulpmill	32.3	6.3	77.5	137	119.0	35.25	225
4. Sirumugai	Viscose	28.7	6.2	30.9	75	20.3	5.8	95

The industrial town of Nagda (Madhya Pradesh) and Kota (Rajasthan) throw their industrial and sewage effluents either directly or indirectly into the river. These effluents in the Chambal river water pollute the water and make it unfit for plankton and fish life which are the main source of food for the people of the region (Bhattacharya and Das, 1981).

Wastewater management consists of (i) treatment of waste-water so as to render it safe, that is, not having harmful effect on the environment, be it land or sea; and (ii) recycling the treated water for reuse if need be. Broadly, sources of wastewater may be divided into two categories -domestic and industrial.

SEWAGE WASTEWATER

Domestic source of waste constitutes mainly sewage. According to Indian Environmental Association, most of the cities in the country discharge untreated sewage directly or indirectly into the coastal system. Interestingly, sewage falls into the category of the most easily treatable effluents, yet most of the sewage generated is untreated. One of the major impediment to having large scale sewage treatment units is the lack of proper sewage collection system beyond the extended suburbs. No doubt, even supplementing these measures in big cities notch a small distance towards surmounting a massive problem, when viewed in the surmounting a massive problem, when viewed in the countryside context. Ray and hope, however, is that it is possible to have fairly effective sewage treatment systems such as "root zone" method which was first used in Isreal in 1968. It was reed bed for the filtration of the waste water. In macrophytes based shallow ponds, the wastewater filtration has a number of advantages which include no operating cost, maintenance free operation, produces almost potable water and has no odour problems.

Recently, impact of sewage waste application on soil strata has been studied by the National Institute of Hydrology, Roorkee. The characterization of sewage indicated that the waste can be used for irrigation purposes. The concentration of various constituents monitored are within normal range and do not constitute a limiting factor for irrigation use. The special distribution of total concentration of various constituents in the soil profile showed that most of the constituents are retained within the top 60-75 cm and no substantial migration takes place in soil strata. The quality of groundwater in the nearby area is quite normal and does not indicate any appreciable sign of contamination due to land application of sewage waste.

INDUSTRIAL WASTEWATER

A number of Indian companies, particularly the larger and medium ones, have woken up to imperative need of having effective waste management/disposal systems. Factors attributing to this are stricter pollution control standards in some states and changing benchmarks for number of companies who aim attaining levels of competitiveness comparable to their international counterparts.

Wastewater management technologies that were developed earlier, focused merely on the treatment and disposal of effluents, that is, end-of-pipe treatment. Waste minimization is relatively new concept in waste management which is now picking up. High water bills are prompting industries to recycle treated water for processes that do not require freshwater. Depending on consumption and cost of water, its percentage in cost of production may vary from 1-5 percent. Industries using relatively large quantities of water include coal, steel, non-ferrous metals, textile weaving and finishing, man-made fibres like rayon, pulp and paper, sugar and distillery units, petroleum refining, petrochemicals, thermal power stations chemicals and fertilizers, drug intermediates, cement and ceremics, Gujarat Fertilizers Corporation near Baroda and Pudumjee pulp and Paper mills Limited, Pune have substantially benefited themselves by extensive recycling of wastewater. Through the recycling of paper machine water alone, the company is reported to be saving Rs. 27 lacks a year energy and disposal expenditure. Although the overall production increased by about 25 percent since 1978, the water requirement was reduced from 11 to 9 MGD. Now by way of water recycle the reduction in overall water consumption and lesser load on demand are considered added advantage, besides reduction in wastewater generation and its treatment. In manufacturing industries in United States, the number of times each cubic metre of water is used has gone up from 1.8 in 1954 to 3.4 in 1978.

TECHNOLOGICAL DEVELOPMENT

The advent of membrane technology represents an important development in the areas to wastewater management. Membrane technology has been found to be very important in food, chemical and pharmaceutical industries. By converting a waste-water stream into the permeate or pure water stream and the concentrate or a reject stream, it enables recovery of some valuable nutrients. Example is cheese making industry where it enables the recovery of proteins value from the whey that is produced. Another area where membrane technology is useful is recovery of oil from industrial oil emulsions containing upto 10 percent oil. However, in case of certain effluents, membrane technology needs to be combined with other available methods of wastewater treatment. Experts feel that use of electrochemical technique, be it for water and effluent treatment or for desalination of water, has certain specific advantages as it does not add to the mineralisation of water or result in secondary pollution. In India,

Central Electrochemical Research Institute has been carrying R & D to develop this method for the purpose of effluent treatment and to solve the environmental problems related to electrochemical industries.

Reasonably advanced indigenous technology is available in waste water management which works out cheaper also as compared to imported ones. Submerged Aerated Fixed Film (SAFF) reactor introduced by Thermax in place of activated sludge process has certain significant advantages. It has lower power requirements, needs less maintenance and has lower operating costs. Overall volume/size of the unit at secondary stage of water treatment. Wastewater management consists of three stages: primary, secondary and tertiary. Tertiary stage is not required if recycling is not to be carried out. Recycling becomes viable only if the cost of recycling the water is less than the cost of purchasing water. However, emerging scenario of scarcity of water is prompting for development of cheaper and appropriate further technologies whereby it is possible to recover and recycle some of the other effluents present in wastewater.

VIABILITY ASPECTS

As a thumb rule, wastewater management is considered economically viable if payback period is less than three years. However, certain factors have deterred a comprehensive waste-water management effort so far. While most large industrial units have a wastewater management policy, the smaller units circumvent the environmental regulations by citing their inability to afford the treatment units. These factors give birth to the idea of setting up of common effluent treatment plants (CSTPs) by a group of industries in industrial areas. In Maharashtra faw such CETPs are operating successfully for the past few years. Situation, however, is not so every where. One of the fundamental problems that deters the smooth and satisfactory functioning of smaller CSTPs is the varying quantity of effluents with fluctuating levels of toxicity due to inconsistency in water treatment by the constituent industries. Efficacy of this system, therefore, depends on the proper management of the treatment plant by ensuring that all industrial units comply with the waste water requirement. Problem thus, is not one of legislation but of enforcement.

In 1996, Maharashtra Industrial Development Corporation (MIDC) adopted a zero discharge policy in 13 industrial areas with an objective to reuse the entire amount of wastewater generated after treatment. Though unsuitable for food processing and pharmaceuticals, such water is found suitable for host of uses such as for gardening and floor washing etc. Pilot plant studies are being undertaken in these industrial areas and project work is expected to be completed by the year 2000. To ensure compliance, non-cooperative industrial units are proposed to be penalised through imposition of charge.

As industrial/domestic demand for water grows, the challenges would be in terms of improving efficient use of water in the years to come. It is being increasingly felt by international forum of experts that we must handle our supplies of water, a scarce and precious resource, more rationally and more cautiously for sustainable development. Hence, there is growing need for effective wastewater management.

CONTROL OF THERMAL POLLUTION

Heat must be removed from condenser cooling waters prior to their disposal into the water bodies. The major principles involved in heat loss are conduction, convection, radiation and evaporation:

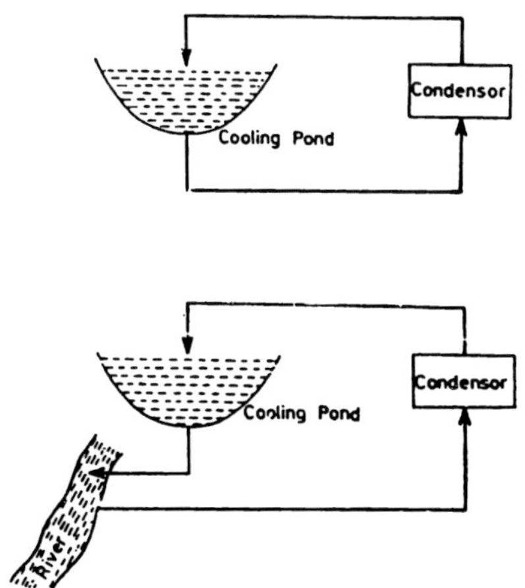

Fig. 59. Cooling ponds for dissipation of heat.

(a) Cooling ponds

The general principle of cooling ponds have been shown in Figure 59. The water from condensers is stored in earth like ponds where natural evaporation brings down the temperature. The water cooling is recirculated or discharged to the nearby water body.

(b) Spray ponds

In spray ponds, the water is sprayed in the cooling ponds with the help of nozzals to convert it into fine droplets which provide more surface area to facilitate efficient heat transfer to the atmosphere (Figure 60).

(c) Wet cooling towers

In wet cooling towers, the heated water is brought in direct contact with continously flowing air. The evaporation brings down the temperature. To increase the surface areas of contact, the water is broken down into droplets by use of spray nozzles or by

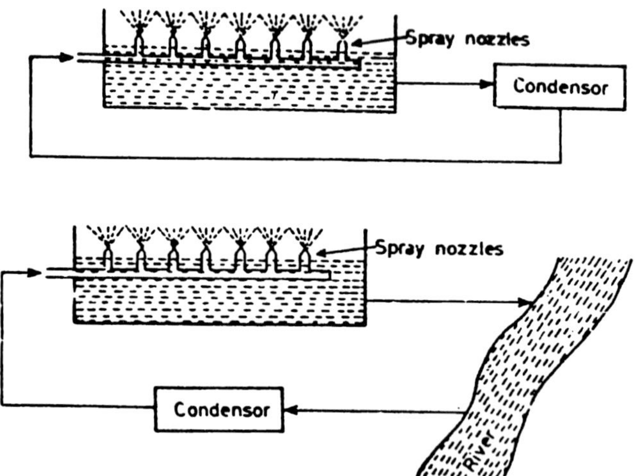

Fig. 60. Spray ponds for dissipation of heat.

splashing it on the packings or baffles in the cooling towers. Principles of wet type of cooling tower have been shown in Figure 61. In atmospheric cooling towers, the heated water falls and the air enters into the tower from the sides and finally comes out after taking the heat from the water. In mechanical draft towers, the use of a fan is made to force the air into the cooling towers. The natural draft towers are hyperbolic in shape and usually very long. The air enters from the lower end of the tower and absorbs heat and water vapours which reduces its density and cause it to rise through the tower maintaining a continuous flow.

(d) Dry cooling towers

In these systems, the air does not come in direct contact with the water, and there is no evaporation. The water circulates in the cooling coils which are cooled with the air flow. The cooling takes place mainly by conduction and convection of heat to the air (Figure 61).

DEFLUORIDATION

The technical method for the removal of excessive fluoridc at domestic and community level using precipitation by aluminium salts is popularly known as "Nalgonda technique" (NEERI, 1987).The technique is adopted when:

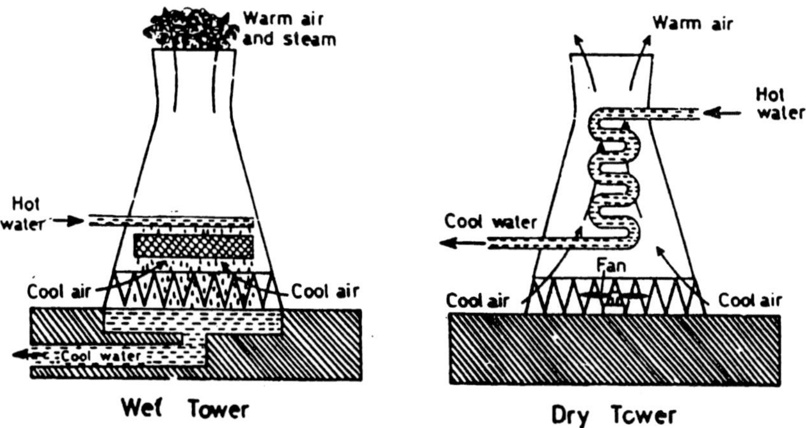

Fig. 61. Wet and dry type of cooling towers.

(a) Low fluoride source and lack of acceptable alternatives.
(b) Dissolved solids are below 1500 mg/l.
(c) Total hardness is below 250 mg/l.
(d) Alkalinity of the water is sufficient.
(e) Raw water fluorides range from 2-20 mg/l.

Defluoridation can be achieved using either of the following sequence of Nalgonda technique. Both these schemes can be used for domestic as well as community water supply.

PRECIPITATION, SETTLING AND FILTRATION

Treatment can be carried out in a bucket of 60 litre capacity with a tap 3-5 cm above the bottom of the bucket for the withdrawal of treated water after precipitation and settling. The raw water taken in the bucket is mixed with adequate amount of lime or Sodium carbonate. Bleaching powder and Aluminium sulphate solution, depending upon its alkalinity and fluoride content. Lime or Sodium carbonate solution is added first and mixed well with water. Alum solution is than added and the water stirred slowly for 10 minutes and allowed to settle for nearly one hour and is withdrawn. The supernatant which contains permissible amount of fluoride is withdrawn through the tap for consumption. The settled sludge is discarded.

PRECIPITATION, FLOATATION AND FILTRATION

Domestic treatment is achieved by using a 100 litre capacity bath type dissolved air floatation cell with hand operated pressure pump. The pump and cell form a compact dissolved air floatation defluoridation system.

Raw water in the cell is mixed with alkali and Aluminium salts. A small quantity of air - water mix from the pressure pump is allowed into the cell. The precipitate with fluoride lifts to the top and floats. The treated water in collected in a bucket filtered through a sand filter, using this cell 100 litre water is available for use in 20 minutes times.

LAND TREATMENT OF WASTEWATER

Land treatment has been defined as the controlled and regulated application of partially clarified wastewater to the land surface to achieve a designed degree of treatment. Sewage has been used on land over the world for several decades as a method of its final disposal. Studies on the feasibility of wastewater reuse on land as a method of treatment for pollution control has been advocated. The advantages of this approach are:

1. The water component and nutrient component of wastewater are recycled through crop irrigation.
2. Pollution of water bodies can be eliminated, and
3. Increase in economic benefits in cases of high crop yield.

The prime objective of sewage farming is to achieve the purification of wastewater for control of water pollution simultaneously with the cultivation at low-cost in comparison to the conventional treatment methods. The mechanism involved in the purification include complex physical, chemical and biological interactions operating together. The physical factors include filtration, absorption, adsorption, photo-oxidation, evapotranspiration, osmosis etc. The chemical factors are ion-exchange, precipitation, solubilisation, oxidation, reduction etc. The biological factors include carbon assimilation, proteolysis, enzymatic hydrolysis, respiration, catalytic decomposition, ammonification, nitrification, denitrification, phosphate mobilisation etc. (Lochr, 1979). The successful operation and efficient management of wastewater reuse for crop irrigation depends on the effluent quality (volume, organic load, nutrient content, total suspended solids, salinity, pH, pathogens etc.), soil properties (texture, permeability, sub-soil drainage, water table depth, cation exchange capacity; exchangeable sodium percentage etc.), and nature of crop species such as tolerance capacity, nutrient and oxygen uptake requirement, duration etc. (Mishra, 1986).

Land application of wastewater is very versatile, efficient and is considered as an advanced process. This may also help in the improvement in physico-chemical and biological quality of soil in wasteland. Wastewater can be disposed of on land for growing suitable plants. This method is not only economical compared to conventional treatment plants but also helps in greening the countryside and thereby improving the quality of life in the urban and rural areas (Das and Kaul, 1992).

Chaturvedi (1985, 1986) found the suitability of a number of tree species in Uttar Pradesh when raised with tannery and dairy wastewater. Observations on survival and growth performance of twelve tree species (Table 141) to tannery wastewater irrigation. Chaturvedi (1986) recorded high survival percentage in eleven species. In respect of biomass *A. nilotica, T. arjuna, C. dichotoma, C. auriculata* and *A.vasica* showed better growth of diameter at breast height when irrigated with tannery wastewater than with ordinary water.

Table 141. Growth in diameter of twelve species on application of tannery wastewater at Kanpur (Chaturvedi, 1986).

Species	Diameter at breast height(cm)	
	Treated	Control
1. *Acacia nilotica* var *indica*	46.60	20.40
2. *Terminalia arjuna*	27.50	16.10
3. *Ficus religiosa*	12.20	8.06
4. *Cordia dichotoma*	27.90	42.50
5. *Haplophragma adenophyllum*	11.80	11.70
6. *Azadirachta indica*	7.55	2.80
7. *Pongamia pinnata*	6.04	2.56
8. *Syzygium cumini*	8.60	6.35
9. *Cassia auriculata*	3.30	1.62
10. *Ficus benghalensis*	3.07	4.61
11. *Adhatoda vasica*	0.40	0.55
12. *Pinus roxburqhii*	0.19	0.32

Agarwal and Kumar (1990) showed that germination and seedling growth of 15 plant species with textile wastewater exhibited 100 percent survival and growth without any detrimental effect for nine species - *Eucalyptus camaldulensis, Azadirachta indica, Prosopis chinensis, Prosopis cineraria,* and *tecomella undulata*. Chaturvedi (1985) observed that *Syzygium cumini* attained better height within one-half year after planting when irrigated with dairy wastewater as compared to plants irrigated with clean tubewell water. Diluted papermill wastewater (50%) showed germination efficiency and growth of paddy and maize comparable to that of control, but pure wastewater inhibited the germination but favoured growth and crop production (Table 142 and 143).

Table 142.. Effect of paper mill effluents on germination growth and crop production in maize (Mishra and Saha, 1989).

Sl. No.	Parameters	Control	50%	100%
1.	Percent germination	73.33	70.0 (-4.54)	43.33 (-40.9)
2.	Plant height (cm)			
	(a) on 31st day	50.25	40.75 (-18.19)	63.5 (26.37)
	(b) on harvest day	142.5	149.8 (5.12)	156.3 (9.68)
3.	Total number of leaves on 31st day	30	27	34
4.	Pigment contents in leaves on 55th day (μg/g fresh weight)			
	Chlorophyll a	620.4	883.08 (+42.0)	1168.21 (+88.0)
	Chlorophyll b	452.6	723.2 (59.79)	739.33 (63.35)
	Carotenoid	0.27	0.48 (77.78)	0.61 (125.90)
5.	Dry weight of shoot (gm) after harvest	42.1	56.25 (33.5)	53.92 (27.98)
6.	Dry weight of ripe cobs (gm) after harvest	11.53	17.25 (49.6)	5.1 (-55.77)

Figures in parentheses show percent increase or decrease over control.

Table 143. Effect *of* paper mill effluent on paddy (Mishra and Sana, 1989).

Sl. No.	Parameters	Control	50%	100%
1.	Pigment content on 60th day (mg/g fresh wt.)			
	Chl. a	903.7	1256.78 (39.07)	1196.07 (32.35)
	Chl. b	589.88	833.48 (41.3)	688.18 (16.7)
	Carotenoid	0.432	0.626 (44.9)	0.512 (18.5)
2.	Number of tillers	31	31 (0)	30 (-32.)
3.	Weight of hundred grains (gm)	1.973	1.911 (-3.14)	1.879 (-4.76)
4.	Average weight of grains produced (gm)	23.15	34.12 (47.39)	34.4 (48.60)
5.	Average shoot length (cm)	67.83	66.5 (-1.96)	77.0 (2.17)
6.	Average dry shoot weight	30.87	30.97 (0.32)	34.76 (12.6)

Figures in parentheses show percent increase or decrease over control.

Sewage and industrial effluents are rich in minerals and sometimes contains some toxic heavy metals. Raw sewage contains toxic elements and heavy metals such as iron, zinc, manganese, lead, cadmium, nickel, cobalt, chromium etc. Though many of are essential plant nutrients, almost all become phytotoxic at higher concentrations. The study of post-irrigation effect on the sewage sludge fed soils showed, increased nickel, chromium and lead contents (Sree Ramulu, 1990). Studies carried out at Kolkata showed high concentration of iron, copper, zinc and lead in parts of vegetation like gourd, mustard, spinach, cauliflower and raddish. In most cases accumulation of lead was beyond the critical limits and thus unfit for human consumption. Although vegetables, crops and fish produced with sewage effluents around Kolkata for decades have not yet been connected with specific health hazards but the risk of heavy metals and other undesirable elements getting into the food chain cannot be ignored (Das and Kaul, 1992).

Somasekhar et al (1984) observed increased pH, organic carbon, sodium, potassium and phosphate content but decreased water holding capacity in soil after prolonged wastewater treatment. It is obvious that prolonged application of sewage and/ or industrial effluents which is rich in nutrients, will tend to increae the soil nutrients. The most important aspect is to utilize the wastewater after primary and secondary treatment thus removing suspended solids and organic matter load respectively. The wastewater containing high levels of toxic heavy metals should not be used in irrigation. Further, human pathogens (bacteria) or toxin producing bacteria spread through use of sewage as an irrigation supplement (Gangawane, 1989).

BIOLOGICAL TREATMENT OF WASTEWATER

Algae, hydrophytes and aquatic macrophytes have been successfully employed to treat wastwaters. The use of algae in wastewater treatment could prove beneficial in different ways since they will bring about oxygenation and mineralization. Algal-bacterial symbiosis has long been proved to be an inexpensive efficient method for reclamation of wastewater over bacterial action alone. Attempts are being made to locate new algae with high growth rate, higher biomass yield and high utilization potential which could be mass cultured in wastewater. Blue-green algae are ideally suited to perform these functions by virtue of their high flexibility to adopt to varied environments and their known nutritional and fertilizer value (Uma and Subramanian, 1990).

Uma and Subramanian (1990) have been successful in removing 25 to 50 percent of the total dissolved solids, calcium and chloride load from wastewater of Ossein industry using *Halobacterium* sp and *Oscillatoria* sp. *O. pseudogeminata* in combination with natural microbial population was efficient remover of nitrates, nitrites and other nutrients in paper mill effluents (Manoharan and Subramanian, 1992). Cyanobacteria have been observed to remove large amount of phosphorus from wastewater (Chan et al. 1979; Tarn and Wong, 1989).

Table 144. Changes in Physiochemical Characteristics in wastewaters during water hyacinth treatment

Parameter	Alluminium Industry				Thermal Power				Fertilizer Industry			
	Day				Day				Day			
	0	10	40	60	0	10	40	60	0	10	40	60
pH	9.2	8.0	83	81	9.0	8.8	8.7	8.9	8.0	8.4	8.4	80
Conductivity (mho cm^{-1})	450	354	378	309	280	230	215	276	1460	1050	512	384
Total dissolved solids	131	122	97	87	220	148	120	109	210	211	205	90
Chemical oxygen demand	–	–	–	–	40	31	22	13	380	300	270	225
Biological oxygen demand	–	–	–	–	07	5	2	0	38	28	28	13
Flouride	15.6	15	9.4	10.5	–	–	–	–	–	–	–	
Total Nitrogen	–	–	–	–	–	–	–	–	167	136	116	112
Total NO_3–N	–	–	–	–	–	–	–	–	14	9.7	6.5	5.9

AQUATIC HYDROPHYTES WASTEWATER TREATMENT

Aquatic hydrophytes are being used in the treatment of wastewaters. The floating water hyacinth *(Eichhornia crassipes)* when grown in high nutrient waters can double their surface area coverage in two weeks (Penfound, 1956). Its easy availability, cultivability, rapid growth rate, high mineral uptake and accumulation, easy harvesting, economic potential as feed, fertilizer and for generation of biogas makes it a suitable species for wastewater treatment.

Mishra et al (1991) observed that a decrease in conductivity (74%), total dissolved solids (57.2%), BOD (67%), COD (50%)., total nitrogen (25%), nitrate nitrogen (67%) in fertilizer factory wastewater during 60 days of water hyacinth treatment. The reduction in conductivity and total dissolved solids inAluminium factory effluents and Thermal Power wastewater was equally high. Fluoride reduction was 38 percent in Aluminium factory wastewater (Table 144).

Table 145. Percentage reduction in some heavy metal content from waste water during water hyacinth treatment

Days of treatment		Percentage reduction				
		Alluminium	Copper	Chromium	Iron	Zinc
1	2	3	4	5	6	7
0		–	–	–	–	–
3	AI	0	0	10	0	10
	TPP	0	67	0	3	11
	FI	–	–	–	17	0
10	AI	5.5	18	6	10	32.5
	TPP	12.5	67	40 1	2	56
	FI	–	100	12	17	20
17	AI	5.5	41	30.5	23	32.5
	TPP	50	67	58.0	67	48.0
	FI	–	100	30	16	70
25	AI	25	41	48.5	80	57.5
	TPP	42	67	66.0	88	63
	FI	–	100	6	54	–

1	2	3	4	5	6	7
32	AI	28.6	59	48.5	83	75.0
	TPP	44	67	66	90	63
	IT	–	100	12	92	74
40	AI	28.6	100	70	87	75.0
	TPP	73	67	74	95	74.0
	FI	–	100	41	93	75.0
48	AI	5.5	100	70	97	75.0
	TPP	75	67	80	95	74
	FI	–	100	53	96	87.5
55	AI	48	100	70	97	75.0
	TPP	75	77	80	95	74
	FI	–	100	70.5	96	87.5
60	AI	48	100	70	97	75.0
	TPP	75	77	80	95	–
	FI	–	100	70.5	96	–

AI - Aluminium Industry TPP - Thermal Power Plant FI - Fertilizer industry.

Water-hyacinth has been successful in reducing BOD and COD significantly (BOD 94% in sugar factory effluents, BOD 84% and COD 88% in dairy effluents, BOD 58% and COD 44% in Pulp and Paper mill effluents, BOD 35% and COD 70% in tannery effluents, COD 92.97% in Textile industry effluents). They are also efficent in removing nitrogen, phosphorus, phenols, pesticides etc (Trivedi and Joag, 1989).

Observations (Table 145) show that waterhyacinth reduces metal content from wastewater. The reduction was 78, 80, and 71.6 percent for Aluminium; 100, 96, 99.6 percent for Iron; 75, 74, 88 percent for zinc; and 100, 76.7 and 100 percent for Copper in Aluminium, Thermal power generation station, and Fertilizer factory wastewater respectively.

BIBLIOGRAPHY

Adler, P. et al. 1970. *Fluorides and human health.* WHO, Geneva.

Adoni, A.D. 1975. Studies on microbiology of Sagar lake. Ph.D. Thesis, Saugar University, Sagar.

Agarwal, H.C. 1983. Pesticide pollution of waters. *In Water pollution and management.* Ed.C.K.Varshney, Wiley Eastern Ltd, New Delhi. 61-73.

Agarwal, R., and Agarwal, S.K. 1990. Physico-chemical characteristics of Kota saree printing effluents, its influence on seed germination and seedling growth of *Cyamopsis tetragonoloba. Acta Ecol.* 12:2:112-118.

Agarwal, R.K., and Kumar, P. 1990. Textile industrial effluents: implications and possible use of afforestation in western Rajasthan. *Ann.Arid* Zone. 29:4:295-302.

Agarwal, S.K. 1982. Eutrophication of Chambal river at Kota. *Acta Ecol.* 4:2:5-9.

Agarwal, S.K. 1983. Water quality of sewage drains entering Chambal river at Kota. *Acta Ecol.* 5:2: 26-29.

Agarwal, S.K. 1986. Ecology of Chambal river at Kota. *Acta Ecol.* 8:1: 13-19.

Aggarwal, S.K., Singh, G., and Sethi, I. 1993. *The degrading environment; causes of global concern.* Commonwealth Publications, New Delhi.

Alabaster, J.S., and Lloyd, R. 1980. *Water quality criteria for freshwater* fish. Buttorworth, London.

Alone, V.R. 1988. Environmental pollution - A case study. *In Recent advances in ecology and environment.* Ed. V.B.Purohit. Today and Tomorrow, New Delhi. 93-102.

Ambasht, R.S. 1981. Responses of aquatic plants to pollution. *Cent. Bd.Prev.Contr. Water Pollut.* Osmania University, Hyderabad. 61-66.

Anderson, R.R. 1969. Temperature and rooted aquatic plants. *Chesapeake Sci.* 10: 157-164.

Arceivala, S.J. 1981. *Wastewater treatment and disposal.* Marcel Dekker Inc. New York.

Atkins, W.R., and Harris, G.T. 1524. *Sci. Proc. R.Dublin Soc.* 18: 1-21.

Baalsrud, K., and Balmer, P. 1973. What ameliorating measures are required in addition to the primary and biological treatment applied in waste water treatment plants. *J. Inform. Bull.* 20: Eur. Fed. protec. *Waters,* Zurich. 76-79.

Baker, J.M. 1970. The effect of oil on plants. *Environ. Poll.*1: 27-44.

Beeton, A. 1969. Changes in the environment and biota of the Great lakes. In *Eutrophication causes, consequences and correctives.* OECD, USA.

Belter, W.G. 1975. Management of waste heat at nuclear power stations, its possible impact on the environment, and possibilities of its economic use. in *Environmental effects* of *cooling systems at nuclear power plants.* International Atomic Energy Agency, Vienna. 1-23.

Benin, A., Navrot, J., Noi, U., and Yoles, D. 1981. Accumulation of heavy metals in aridzone soils irrigated with treated sewage effluents and their uptake by Rhodes grass, *J. Environmen. Qual.* 10:4:536.

Bernard, R.P. 1965. J.Hugy. London. 63: 537-563.

Berrow, M.L., and Webber, J. 1972. Trace elements in sludges. *J.Sci.Pd.Ajric.* 23: 93-100.

Best, E.P.H. 1979. Growth substances and dormancy in *Ceratophyllum demersun. Physiol. Plant.* 45: 399-406.

Bharti, S.G., and Murthy, S.R.K. 1990. Effect of industrial effluents on river Kali around Dandeli, Karnataka Part I. Physico-chemical complexes. *Indian J.Environ. Hlth.* 32:2:167-172.

Bhattacharya, A.K., and Das, R.R. 1981. Industrial pollution in Madhya pradesh A brief review. *ICEE,* New Delhi. 73-74.

Bhoota, B.V. 1975. *Commerce,* 131: 3372: 157-162.

Black, A.P., and Goodson, J.H. 1952. The oxidation of sulphides by chlorine in dilute aquous solutions. *J.Amer.Water Works Assoc.* 44: 309.

Boralkar, D.M., Trivedi, R.K., and Kulkarni, A.Y. 1981. Studies on salt build-up in the Krishna river, Maharashtra state. *Acta Limnol.Indica.* 1: 5-9.

Borneff, J., and Kunte, H. 1964. *Arch. Hyg. Berl.* 148: 585-597.

Bradshaw, J.S. et al. 1972. Seasonal variations in residues of chlorinated hydrocarbons pesticides in the water of Utah lake drainage system 1970-71. *Pest. Mon.J.* 6: 166.

Bricker, O.P., and Garrels, R.M. 1967. Mineralogical factors in natural water equilibria. In *Principles and application of water chemistry.* Ed. S.D.Faust and J.V.Hunter. Wiley, New York.

Brisou, J. 1968. *Bull. World Hlth. Org.* 38: 79-118.

Brown, V.M., Shurben, D.G., and Shaw, D. 1980. Studies on water quality and the absence of fish from some polluted English rivers. *Water Res.* 4: 363-382,

Budyko, M.L. 1930. *Global ecology.* Progress Publishers, Moscow.

Bulusu, K.R., Nawalakha, W.G., Kulkarni, D.M., and Litada, S.L. 1985. Fluoride: Its incidence in natural water and defluoridation methods. *Silver Jubilee Commemorative Volume,* NEERI, Nagpur.

Buswell, A.M., and Larson, T.E. 1939. 2. *Amer. Water Works Assoc.* 29: 1978.

Cairncross,S. 1990. Water supply and the urban poor. In *The poor die young; housing and health in the Third World cities* Eds. J.E.Hardoy, S. Cairncross, D. Satterthwaite, Earthscan, London, 109-126.

Canter, L.W. 1977. *Environmental impact assessment.* McGraw Hill Book Co, New York.

Carrington, E.G. 1978. The contribution of sewage sludge to the dissemination of pathogenic micro-organisms in the environment. *Water Research Centre Technical* Report.TR. 7.

Carrol, D. 1962. Rainwater as a chemical agent of geological progress- A review. *Geological Survey Water Supply,* 1535-6.

Carter, A. 1973. *Kogai,* 1:1:10-11.

Chakrabarty, R.N., and Gupta, S.K. 1980. A low cost sewage treatment plant for control of Jamuna river pollution, *IAWPC Tech.Annual,* 6 & 7: 45-58.

Chan, K.Y., Wong, K.H., and Wong, P.K. 1979. Nitrogen and phosphorus removal from sewage effluents with high salinity by *Chlorella salina. Environ, Poll.* 18: 139-146.

Chang, S.L. 1968. Bull. *World Hlth.Org.* 28: 410-414.

Chakrabarti, C., and Chakrabarti, T. 1988. Effect of irrigation with raw and differentially diluted sewage and application of primary settled sewage sludge on wheat plant growth, crop yield, enzymatic changes and trace elements uptake. *Environ. Poll.* 52: 219-235.

Chaney, R.L. 1973. In *Recycling municipal sludge and effluents on land.* National Association of State Universities of Land Grant College, Washington DC. 129-141.

Chapman, H.D. 1966. In *Diagnostic criteria for plant and soils.* University of California, Riverside. 484-499.

Chaturvedi, M.C. 1975. Water in India's future. *Commerce.* 131:3372: 15-18.

Chaturvedi, A.N. 1985. Jamun plantations for controlling dairy effluents. J.Trop.For. 1:3:210-274.

Chaturvedi, A.N. 1986. Trees and shrubs for control of tannery wastewater in India. *Environ. Conserv.* 13: 2: 164-165.

Chauhan, A.K. 1990. Distillery and refinery waste characteristics, recovery and its impact on aquatic environment with reference to periphyton. *IAEM Annual,* NEERI, Nagpur.

Chauhan, E. 1977. Quality of human environment in Indian cities -. Delhi. In *Current trends in Indian environment.* Eds Desh Bandhu and E. Chauhan. Today and Tomorrow, New Delhi. 79-92.

Cheng, C.I., and Blackwell, R.O. 1968. *Amer. J.Epidem.* 88:7-24.

Chhapgar, B.F. 1993. Oil and water do not mix. *The Hindu Survey* of *the environment.* Ed. N.Ravi, The Hindu, Madras. 62-67.

Choubisa, S.L., Choubisa, L., and Choubisa, D.K. 2001. Endemic fluorosis in Rajasthan. *Indian J.Environ. Hlth.* 43:4:177-189.

Clapham, Jr. W.B. 1973. *Natural ecosystems.* The Macmillan Company, New York.

Collingwood, R.W. 1979. The effect of algal growth on the quality of treated water. In *Biological indicators of water quality.* Eds. A.James and L.Evans. John Wiley & Sons, New York.

Connell, D.W. 1981. Petroleum hydrocarbons in the Hudson Raritan estuary. Marine Science Research Centre, State University of New York, Stony Brook, Working paper. 3.

Connell, D.W., and Miller, G.J. 1981 (a). Petroleum hydrocarbons in aquatic ecosystems - behaviour and effects of sublethal concentrations. Part I. *CRC Crit. Rev. Environ. Control,* 11: 37.

Connell, D.W., and Miller, G.J. 1981 (b). Petroleum hydrocarbons in aquatic ecosystems - behaviour and effects of sublethal concentrations. Part II. *CRC Crit Rev. Environ.Control,* 11: 105.

Connell, D.W. and Miller, G.J. 1984. *Chemistry and ecotoxicology* of *pollution.* John Wiley & Sons, New York.

CPCB. 1978-79. *Scheme* for *zoning and classification of Indian rivers, estuaries and coastal waters.* ADSORBS/ Central Pollution Control Board, New Delhi.

CPCB. 1989. *Status of water supply and wastewater collection, treatment and disposal in Class* I *cities.* Central Pollution Control Board, New Delhi.

CPCB. 1995. *Pollution control acts, rules and notifications issued thereunder.* Central Pollution Control Board, New Delhi.

CPHERI. 1972. *Environmental pollution.* Central Public Health Engineering Reasearch Institute, Nagpur.

Dahiya, S., Kaur, A., and Jain, N. 2000. Prevalence of fluorosis among school children in rural area, district Bhiwani -A case study. *Indian J.Environ.Hlth.* 42: 4: 192-195.

Das, D.C., and Kaul, R.S. 1992. *Greening wastelands through wastewater.* National wastelands development board. Ministry of Environment and Forests, New Delhi.

Das, R.R. 1988. Water pollution'. In *Environmental issues and researches* in *India.* Ed. S.K.Agarwal and R.K.Garg. Himanshu Publications, Udaipur. 85-96.

Dayama, O.P. 1997. Influence of dyeing and textile water pollution on Modulation and germination of gram (*Cicer* arietinum). *Acta Ecol.* 9:2:34-37.

Deoras, P.J. 1981. *Indian.J.Environ. Prot.* 1:1:21-25.

DeRoe, H.C. 1980. Nitrate fluctuation in groundwater as influenced by use of fertilizer. Connecticut Agric.Exp. Station, University of Connecticut, New Haven, CT Bull. 779.

Dillenberg, H.O., and Dehnel, M.K. 1960. Toxic waterbloom in Saskatchewan. *Canadian Med.Assoc.Jour.* 83: 1151-1154.

Dix, H.M. 1981. *Environmental pollution.* John Wiley & Sons. London.

Dreisbach, R.H. 1971. *Handbook of poisoning*. Los Altos, California.

Doudoroff, P., and Katz, M. 1953. Critical review of literature on the toxicity of industrial wastes and their components to fish II. the metals and salts. *Sewage Ind. Waters*, 25: 802-839.

Edwards, C.A. 1970. *CRC Crit.Rev.Environ. Contr.* 1: 7-67.

Edwards, C.A. 1973. *Persistent pesticides in the environment*. CRC press Cleveland, Ohio.

Edwards, C.A. 1977. Nature and origin of pollution of aquatic systems by pesticides. In *Pesticides* in *aquatic environments* . Ed. M.A.O.Khan, Plenum Press, New York.

Edwards, P. 1985. *Aquaculture; A component of low cost sanitation technology*. The World Bank, Washington DC.

E.E.C. 1977. *Water purification* in *the* EEC; *A state of the art review*. Pergamon Press.

Ehiri, J.E., Azubuike, M.C., Ubbaonu, C.N., Anyanwu, E.C., Kasimir, M.I., and Ogbonna, M.O. 2001. Critical control points of complementary food preparation and handling in eastern Nigeria. *Bull. World Hlth* Org. 79:5: 423-433.

Ellis, M.M. 1937. Detection and measurement of stream pollution. *Bull. US Fish. Wash.* 22: 365-437.

Erickson, P.A., and Raynolds,J.T. 1971. The ecology of reservoirs. In *Encountering the environment*. Ed. A. Mayer. Von Nostrand Reinhold, New York. 163-171.

Ethrington,J.A. 1975. *Environment and plant ecology*. John Wiley & Sons, London.

Federal Water Quality Administration 1970. *Clean Water for the 1970s*. Washington DC.

Fender, B.S., and Kharat, R.B. 1992. Assessment of toxic metals ions in waste water of Koradi Thermal Power plant. *J.Ind. Poll. Control.* 8:1:21-24.

Fillos, J., and Swanson, W.A. 1975. The release of nutrients from river and lake sediments. *J.Water Poll. Contr.Fed.* 47: 1032-1041.

Fox, C.J.J. 1909. *Trans. Faradey* Soc. 5: 68.

Fruh, G.E. 196t. In *Advances in water quality improvement*. Eds. E.F.Bloyna and W.V. Eckenfelder. University of Texas Press, Austin.

Fruh, G.E., and Lee, G.F. 1968. The effect of eutrophication upon the water resources management of the Yahara river basin. Report of the water chemistry program, university of Wisconsin, Malison, Wisconsin.

Ganapati, S.V. 1960. *Proc.Symp.Algology. ICAR*, New Delhi. 204-218.

Gangawane, L.V. 1989. Sewage farming and micro-organisms- A review. In *Soil pollution and soil organisms*. Ed.P.C.Mishra. Ashish Publishing House, New Delhi. 87-96.

Garcia, M.J., and Page, A.L. 1976. Influence of ionic strength and inorganic complex formation on the sorption of trade amounts of Cadmium by montmorillonites. *J.Soil.Sci. Soc. Amer*. 40: 651-663.

Gautam, A.K., and Agarwal, S.L. 1987. Acute zinc toxicity to freshwater fish: *Cirrhinus mrigala* (Ham). *Acta Ecol*. 9:2:38-43.

Gautam, A.K., and Lal, S.B. 1988. Toxic effects of Copper, cadmium and zinc to a larvicidal fish: *Gambusa affinis Patruclus* (Baird and Girard). *Acta Ecol*. 10:2: 59-63.

Gautam, P.C., Khandekar, C.D., and Agarwal, S.K. *1979*. Toxicity of DDT to *Channa punctata* (B1). *Acta Ecol*. 1; 34-36.

Gehlar, L.W. 1972. The aquous underground. *Tech. Rev*. 75:3-11.

GESAMP. 1977. *Impact of oil on the marine environment*. Joint Group of Experts on the Scientific Aspects of Marine Pollution. Rep. Study: no.6 FAO, Rome.

Ghosh, N.C., and Seth, S.M. 1994. Atmospheric pollutants and their effects on quality of water. *Indian J.Environ. Hlth*. 36: 2: 104-114.

Glenn, L., Bishop, H., and Dunbar, G.E. 1960. Algal parasites in fish. *Prog. Fish. Culturist*. 22: 120.

Golterman, H.L., Clymo, R.S., and Ohnstad, M.A.M. 1978. *Methods for physical and chemical analysis* of *freshwater*. IBP Handbook No. 8. Blackwell Scientific Publications, Australia.

Gopal, R., Bhargava, T.N., Ghosh, P.K., and Rai, S. 1983. Increase of fluoride and nitrate in waters of Banner, Jaisalmer and Bikaner. Trns. *Indian* Soc. *Desert Technol*. 8: 2: 10-12.

Grace, J.B., and Tilly, L.J. 1976. Distribution and abundance of submerged macrophytes including *Myriophylluna spicatum* L. in a reactor cooling reservoir. *Arch. Hydrobiol*. 77:475-485.

Grant, W.D., and Long, P.E. 1981. *Environmental microbiology*. Blackie, Glasgow.

Graves, G.W., Hatfield, G.B., and Whinston, A.B. 1972. Mathematical programming in regional water quality management. *Water Resources Research,* 2:8:273-291.

Gupta, A.K. 1989. Assessment of phenols in petroleum refinery effluents. *Acta Ecol.* 11:1:38-41.

Gupta, S.C. 1981. Evaluation of quality of well waters in Udaipur district. *Indian J.Environ.Hlth.* 23:3:195-202.

Haag, R.W., and Gorham, P.R. 1977. Effect of thermal effluent on standing crop and net production of *Elodea canadensis* and other submerged macrophytes in lake Wabamum, Alberts. *J.Appl.Ecol.* 14:835-851.

Hahne, H.C.H., and Kroontje, W. 1973. Significance of pH and chlorine concentration on behaviour of heavy metal pollutants Mercury, cadmium, zinc and lead. *J.Env. Qual.* 2:444-450.

Hammerton, L. 1972. The Nile river - a case history. In *River ecology and man.* Eds. R.T.Oglesby, C.A.Carlson and J.A. MaCann. Academic press, New York.

Handa, B.K., Goel, D.K., Kumar, A., and Sandhi, T.N. 1982. Pollution of groundwater by nitrate in Uttar Pradesh. *IAWPC Tech. Abn.* 9:95-103.

Hasler, A.D. 1968. Man induced eutrophication of lakes. In *Global effects of environmental pollution.* Ed.S.F.sanger. D. Reidal, Dordrecht. 111-125.

Hayes, F.R., and Phillips, J.E. 1958. *Limnol. Oceanog.* 3:459.

Heaney, S.I. 1971. The toxicity of *Microcystis aeruginosa* Kutz. from some English reservoirs. *Water Treat.Exam.* 20:235-244.

Heise, H.A. 1951. Symptoms of hey fever caused by algae II. *Microcystis* another form of algae producing allergenic reactions. *Ann. Allergy,* 9;100.101.

Holmes, J.W., and Talsma, T. 1981. *Land and stream salinity.* Elsevier Scientific Publications Co. Amsterdam.

Hood, D.W. 1965. In *The Encyclopedia of oceanography.* Ed. R.W. Fairbridge. Van Nostrand, New York.

Hume, N.B., and Gunnerson, C.E. 1962. Characteristics and effects of hyper ion effluents. *J. Water Poll. Contr. Fed.* 34:15.

Hutchinson, G.E. 1957. *A treatise on Limnology.* Vol.1. John Wiley & Sons, New York.

Hutchinson, G.E. 1969. Eutrophication, past and present. In *Eutrophication, causes, consequences, corrections.* National Academy of Science, Washington. 12-26.

ICMR. 1975. Indian Council of Medical Research. Special Report. Series No 44. New Delhi.

Irving, S. 1974. *Industrial pollution.* Non Nostrand, Reinhold Company.

Jain, R.K., and Agarwal, A.K. 1989. Water pollution due to Saree printing industry effluents. *Acta Ecol.* 11:1:50-54.

Jain, R.K., and Kumari, S. 1990. Effect of Saree printing industry effluents on seed germination, seedling growth and total biomass on *Spinacea oleracea. Acta Ecol,* 12:1:19-22.

Jemmer, M.K. 19S7. *Water resources and water management.* Elsvier Amsterdam.

Jernefelt, H. 1925. Zur limnologie einiger Gevasser Finnlands. *Ann. Soc. Zool. Bot. Finnical* Vanamo. 2:185-352.

Jogdand, S.N. 1995. *Environmental biotechnology.* Himalaya Publishing House, New Delhi.

Joshi, M.C., Thukral, A.K., and Chand, R. 1982. *Indian J.Env.Hlth.* 24:4:292-297.

Kanwar, J.S., and Mehta, K.K. 1968. Toxicity of fluoride in some well waters of Haryana and Punjab. *J.Agric.Sci.* 38:881-886.

Kant, S. 1987. In *Environment and ecotoxicology.* Eds. R.C.Dalela Y.N.Sahai and S.Gupta. The Academy of Environmental Biology, Muzaffarnagar. 49-74.

Kasliwal, R.M., and Solomon, S.K. 1974. Fluorosis in a case report. *J.Assoc.Physic. India.* 7: 56-59.

Keegan, R.T., and Leed, J.V. 1970. Dynamic programming for water quality control. *Water Resources Bulletin.* 5:6;235-248.

Kelling, K.A., Keeney, D.R., Walsh, L.M., Ryan,J.A. 1977. *J.Water Qual.* 613:525-528.

Kenner, E.E. 1978. Faecal streptococcus indicators. In *Indicators of viruses in water and food.* Ed.G.Berh. Ann. Arbour Science Publications, MI.

Kerr, P.C., Paris, D.F., and Brockway, D.L. 1970. *The interrelation* of *carbon and phosphorus in regulating heterotrophic and autotrophic populations in aquatic ecosystems.* US Govt. Printing Office, Washington DC.p.53.

Khangrot, B.S. 1980. *J. Ichthyl.* 1:33-36.

Kitamura, S. 1968. Determination of mercury contents in bodies of inhabitants, cats, fishes and shells in Minamata district and the mud of Minamata bay. In *Minamata* disease.Kumamoto University, Japan, 257-266.

Kiyoura, R. 1963. *Intern.J.Air Water Pollut.* 7: 459-470.

Klein, L. 1957. *Aspects of river pollution.* Butterworth Publications, London.

Klein, L. 1973. *River pollution II. Causes and effects.* Butterworth, London.

Kneese, A.V. 1964. *The conomics of regional water quality management.* John Hopkins Press, Baltimore.

Koff, R.S. 1990. *CRC Crit.Review Environ. Contr.* 1: 383-442.

Kopp, J.F. & Kroner, R.C. 1970. *Trace metals in waters of the United States: A 5 year summary* of *trace metals in rivers and lakes of the US.* US Dept of Interior, FWPCA, Division of Pollution Surveillance, Cincinnati, Ohio.

Kshirsafar, S.R. 1968. *Sewage and sewage treatment.* Roorkee Publishing House, Roorkee.

Kudesia, V.P., and Sharma, C.B. 1981. *IAWPC Tech Ann.* 8:168-170.

Kumar, H.D. 1977. *Modern concepts* of *ecology.* Vikas Publishing House, New Delhi.

Kumar, H.D., and Rai, L.C. 1978. Zirconium induced precipitation of phosphate as a means of controlling eutrophication. *Aquatic Botany,* 4: 359-366.

Kuntzel, L.E. 1969. Bacteria, carbon dioxide and algal bloom. *J.Water Pollut. Contr. Fed.* 41: 1737-1747.

Lahiri, A. 1975. Planning water use for 2000 AD. Commerce,131: 3372: 19-27.

Laxmanan, A.R., Rao, T.K., and Wiswanathan, S. 1986. Nitrate and fluoride in drinking waters in the twin cities of Hyderabad and Secunderabad. *Indian J. Environ. Hlth.* 28:1:39-47.

Lee, G.F., and Hoadley, N.W. 1967. Biological activity in relation to the chemical equilibrium composition on natural waters. *Adv. Chem. Ser.* 67. *Amer. Chem. Soc.* 319-339.

Leeper, G.W. 1978. *Managing the heavy metals on the land.* Marcel Dekker, New York.

Leibee, H.C., and Smith, R.L. 1953. Control of algae, a means of prolonging the life of lakes. Water Eng. 24: 620-621.

Lehr, J.H., Tyler, E.G., Wayne, A.P., and Jack, D. 1980. *Domestic water treatment.* McGraw Hill Book Co, New York.

Lenihan, I., and Fletcher, W.W. 1977. Environment and man; *The marine environment.* Blakie, Glasgow and London.

Little, R.J. 1970. Air pollution affecting the performance of domestic animals. Agriculture Handbook No. 38O. ARS, USDA, Washington DC.

Livingston, D.A. 1964. *Chemical composition of rivers and lakes.* US Geol.Surv.Prof. Paper. 440.

Lochr, R.C. 1979. *Land application* of *wastes.* Vol. I and II. Van Nostrand, Reinhold, New York.

Ludwig, H.F., and Storrs, P.N. 1973. Regional water quality management. *J. Water Poll. Contr. Fed.* 10:45:2065-2071.

Lueschow, L.A., Helm, J.M., Winter, D.R., and Karl, G.W. 1970. Tropical nature of selected Wisconsin lakes. *Wisconsin Academy of Sciences, Arts and Letters. 58:* 237.

Lowden, G.F., Saunders, C.L., and Edwards, R.W. 1969. *Water tret. Exam.* 19:275-294.

Lunkad, S.K. 1994. Rising nitrate levels in groundwater and increasing N-fertilizer consumption. *Bhu-Jal news,* 4-10.

Manoharan, C., and Subramanian, G. 1992. Interaction between paper mill effluents and the Cyanobacterium: *Oscillatoria pseudogeminata* var *unigranulata. Poll. Res.* 11:2:73-84.

Mason, C.F. 1981. *Biology* of *freshwater pollution.* Longman, New York.

Mathur, Y.P., and Kumar, P. 1990. Concerns and trends in environmental levels of nitrates. *Indian J. Environ. Hlth.* 32; 2: 97-108.

Mathur, G.M., Tamboli, B.L., Mathur, R.H., Ray, A.K., Mathur, G.C., and Goyal, O.P. 1976. Preliminary epidemiological investigation of fluorosis in Surajpura and Pratappura villages in Sarwar tehsil, Ajmer district (Rajasthan). *Indian J.Present.Secio. Med.*7; 90-93.

Mauria, S. 1986. In *Desert environment; Conservation and management.* Eds. K.A.Shankarnarayan and V.Shankar. CAZRI, Jodhpur. 89-94.

McClure, F.J. 1970. *Water fluoridation,* US Dept of Health, education and welfare. National Institute of Hlth, Bethesda, Md.

McKee, J.E., and Wolf, H.W. 1963. *Water quality criteria*.2nd Edn. Respurce Agency of California. State Water Quality Control Board Publication, 3-A.

McNamara, J.R. 1976. An optimization model for regional water quality management. *Water Resources Research*,12; 125-134.

Mehta, R.S. 1978. *AIWPC* Tech.Ann. 4; 153-156.

Michael, A.D. 1977. The effects of petroleum hydrocarbons on marine populations and communities. In *Fate and effects of petroleum hydrocarbons in marine organisms and ecosystems,* Ed. D.A.Wolfe. Pergamon Press, New York.

Mihursky, J.A. 1969. Thermal loading: New threats to aquatic life. *Catalyst,* 2; 6-9.

Miles, J.R.W., and Harris, C.R. 1971. Insecticides residues in a stream and a controlled drainage system in agricultural areas of south western Ontario. *Pest. Monit.J.* 5; 289.

Miller, G.J. 19S2. Ecotoxicology of petroleum hydrocarbons in the Marine environment. *J.Applied Toxicology,* 2:88.

Miller, S.J., and Connell, D.W. 1980. Occurrence of petroleum hydrocarbons in some Australian seabirds. *Aust. Wildl. Res.* 7: 281.

Mishra, P.C., and Saha, S. 1989. Agro-potentiality of papermill wastewater. In *Soil pollution and soil organisms.* Ed. P.V. Mishra. Ashish Publishing House, New Delhi.88-120.

Mishra, P.C., Patri, M., and Panda, M. 1991. Growth of water hyacinth and its efficiency in the removal of pollution load from industrial waste water. *J. Ecotoxicol. Environ. Monit.* 1:3:218-224.

Mishra, R.P. 1986. Wastewater recycling for pollution control. *Indian Environ. Jour.* 15: 9:25-32.

Mishra, S.G., and Mani, D. 1992. *Metallic pollution.* Ashish Publishing House, New Delhi.Mishra

Mishra, P.C., and Shukla, P.K. 1992. Crop as a means of assessing soil pollution. In *Indian environment.* Ed. p.Singh. Ashish Publishing House, New Delhi. 95-102.

Mohan Rao, G.J. 1971. Water pollution aspects in pulp and paper industry. Paper presented at the annual meeting of Indian pulp and paper technology Association. No 8-9.

Moore, S.F., and Dwyer, R.L. 1974. Effects of oils on marine organisms: A critical assessment of published data. *Water Res.* 8:819.

Mortimer, C.H. 1941-42. The exchange of dissolved substances between mud and water in lakes. *J. Ecol.* 29: 280-329; 30: 147-201.

Moosley, J.W. 1967. Transmission of viral diseases by drinking Water. In *Transmission of viruses by water route*. Ed. G.Berg. Interscience, New York.

Mur, L.R. 1980. Concluding remarks, in *Developments in Hydrobiology.* Ed. J.Barica and L.R. Mur. W. Junk, The Hague.

Murthy, Y.K. 1975. Water for tomorrow. *Commerce.* 131;3372:9-12.

Nag, B.S., and Kathpalia, G.N. 1975. Water resources in India. Proc. *Second World Congress* on *Water Resources.* Report of the Irrigation Commission , Ministry of Irrigation and Power, New Delhi. Vol.2: 373.

Narayan, S.N., and Madhyastha, M.N. 1985. *Poll. Res.* 4:2: 77-79.

Naumann, E. 1919. Nagra synpunkle angaende limnplanktond okologi med sarskild hansyn till fytoplankton. *Sv. Bot. Tidskr.* 13: 129-158.

NEERI, 1975. *Course mannual on industrial waste treatment.* National Environmental Research Institute, Nagpur.

NEERI. 1987. *Technology mission on drinking water in villages and related water management.* National Environmental Engineering Research Institute, Nagpur.

Neff, J.M. 1979. *Polycyclic aromatic hydrocarbons in the aquatic environment.* Allied Science Publishers, London.

Nemerow, N.L. 1978. *Industrial water pollution: Origin, characteristics and treatment.* Addison Wesley Publishing Co, Inc. Philippines.

Newton, M.E., and Fetteroff, C.M. 1966. *Limnological data from the lakes, Genesee and Livingston counties, Michigan,* September 1965. Water Resources Commission, Bureau of Water management, Michigan Depptt of Natural Resources, lansing, 16 p.40 Appendices.

Niyomkha, P.A., and Wonghanchao, W. 1975. Pollution in Thailand: A survey. *Kogai.* 34-65.

Nygaard, G. 1949. Hydrobiological studies on some Danish ponds and lakes. Part II. The quotient hypothesis and some new or little known phytoplankton organisms. *Kongel Danske Vidensk Selskab Biol. Skrift.* 7: 293.

Oborn, E.T., and Higginson, E.C. 1954. *Biological corrosion* of *concrete, joint report.* Field crops Research Branch. Agric. Res. Service USDA and Bur. Reclmation, US Dept of Interior. p.8.

Ohle, W. 1934. Chemische and physikoliache water suchungen norddeutscher seen. *Arch. Hydrobiol.* 36: 356-364.

Okun, D.A. 1977. *Regionalization of water management in England and Wales.* Applied Science publishers, London.

Olaniya, M.S., and Sacena, K.L. 1977. *Indian J.Environ. Hlth.* 19; 3: 176-188.

Pachpurkar, S.K. 1976. Water pollution in metropolitan cities with special reference to Bombay metropolis. *J.Inst.Town Planners, India.* 88: 88-89.

Palmer, C.M. 1980. *Algae and water pollution.* Castle Housing Publication, Ltd.

Pande, Y.N., and Patel, K.K. 1997. Pollution tolerant macrophytes of river Terhi at Nawabganj industrial site of district Gonda. *Acta* Ecol. 19: 1: 23-26.

Pandey, S.P., Narayanswamy, V.S., and Hasan, M.Z. 1979. Quality of well waters in Nagpur with regard to nitrate, nitrites: Phase I. *Indian J.Environ. Hlth.* 21: 35-46.

Parker, B.W. 1965. Minutes of 3rd Annual Conference Patchment Estuary studies, University of Maryland, CBL. Ref. No 65-23.

Parker, F.L., and Krenkel, P.A. 1969, *Thermal pollution: State of art.* Report No. 3. Vanderbilt University.

Parmasivan, M., and Sreenivasan, A. 1981. Changes is algal flora due to pollution of Cauvery river, *Indian J.Environ. Hlth.* 23: 222-239.

Paul, M. 1995. Sea change: Marine alert. *Sanctuary,* 15: 1:39-45.

Pearsall, W .H. 1921. The development of vegetation in the English lakes considered in relation to the general evolution in glacial lakes and rocks basins. *Proc. Royal Soc.* 92: 259-284.

Penfound, W.T. 1956. Primary production of vascular aquatic plants. *Limnol. Oceanogr.* 1: 92-101.

Pervej, S., and Pandey, G.S. 1994. Contamination of river water and sediments by thermal power ashpond discharge. *Indian J. Environ. Hlth.* 36: 1: 8-12.

Phillips, J.E. 1964. In *Principles and applications of aquatic microbiology.* Ed. H.Henkelektan and N.C.Dondero. Wiley, New York.

Pickering, O.H., and Henderson, C. 1964. *Proc. Purdue University Industrial Waste Conference,* 578-591.

Pillai, M.K.K., Agarwal, H.C., and Yadav, D.V. 1977. Tolerance, uptake and metabolism of DDT in *Gambusa affinis*. *Indian J. Expt. Biol.* 15: 40.

Poole, N.J., Wildish, D.J., and Kristmanson, D.D. 1973. The effect of pulp and paper industry on the aquatic environment. *CRC Crit. Rev. Environ. Contr.* 8: 153.

Prescott, G.W. 1939. Some relationship of phytoplankton in limnology and aquatic biology. *Pub. Amer. Assoc. Ad. Sci.* 10: 65-78.

Price, D.R.H. 1979. Fish as indicators of water quality. In *Biological indicators of water quality*. Eds. A.James. and I.Evisoa. John Wiley & Sons, New York.

Purohit, S.S. 1988. Fluoride toxicity and plant life. In *Environmental issues and researches in* India. Eds. S.K. Agarwala, R.K.Garg, Himanshu Publications, Udaipur. 255-292.

Raj, M. 1977. Water, sewarage, health and sanitation levels in urban areas, in *Current trends in Indian environment*. Eds. Desh Bandhu and E.Chauhan. Today and Tomorrow, New Delhi. 99-106.

Rajannan, G., and Kandasamy, D. 1990. Environmental pollution of Pykara river by protein industry effluents. *Indian J. Environ. Hlth.* 32:3:230-237.

Ramamohan, Rao, N.V., and Bhaskaran, C.S. 1964. *Indian J.Med. Res.* 52: 180-186.

Ranga, M.M., and Yadav, B.L. 1989. Problems of eutrophication in lakes of Bhilwara. In *Wetland conservation*. Eds L.N. Vyas and R.K.Garf. Environment Community Centre, Udaipur. 227-231.

Rao, K.L. 1975. *India's water wealth*. Orient Longman Ltd, New Delhi.

Rao, K.L. 1979. *India's water wealth*. Orient Longman, New Delhi.

Rao, S. 1981. State of art of sewage treatment in India. Lecture delivered in International Conference on role of youth in environmental conservation, Karad, March 1981.

Rao, S.N., Chaubey, R., and Srinivasan, K.V. 1990. Ganga water quality in Bihar. *Indian J.Environ. Hlth.* 32:4:393-400.

Rawson, D.S. 1956. Algal indicators of trophic lakes types. *Limnol. Ocoanogr.* 1;18-25.

Rawson, M.N. 1960. Limnological comparison of twelve large lakes in northern Saskatchewan, *Limnol.* Oceanogr.5:195-211.

Rice, S.D., Short, J.W., and Karinen, J.F. 1977. Comparative oil toxicity and comparative animal sensitivity. In *Fate and effects of petroleum hydrocarbons in marine orgânisms and ecosystems.* Ed. D.A.Wolfe. Pergmon Press, New York.

Roberts, G.H., Grihiey, J., and William, E.H. 1940. *Chemical methods for the study of river pollution.* London.

Robertson, H.E., and Riddle, W.A. 1949. *Canad, J.Publ. Hlth.* 40: 72-77.

Rodhe, W. 1969. Crystallization of eutrophication concepts in northern Europe. In *Eutrophication causes, consequences, correctives.* National Academy of Science, Washington DC. 50-64.

Ross, F.F. 1977. Summary review of the effects of thermal discharges into fish water. In *Biological balance and thermal modifications.* Ed. M.Marous. Oxford, New York. 109-177.

Rowe, D.R., Canter, L.W., Snyder, P.C., and Mason, J.W. 1971. Dieldrin and endrin concentration in a louisiana estuary. *Pest. Monito. J.* 4: 177.

Ruttner, F. 1937. Int. *Revue ges hydrobiol. Hydrogr.* 35:7-34.

Sampath, V., Sreenivasan, A., and Ananthanarayanan, R. 1981. Mollascs as indicators of organic enrichment and pollution in the Cauvery river system. *Cent. Bd. Prev. Contr. Water Pollut.* Osmania University, Hyderabad. 149-162.

Sastry, C.A. 1986. In *Pollution control handbook.* Ed. P.L. Oiwakar Rao, Utility publications, Secunderabad. 39-48.

Sawyer, C.N., and McCarty, P.L. 1978. *Chemistry for environmental engineering.* McGraw Hill Book Co. Tokyo.

Saxena, R., and Nathawat, G.S. 1989. Fluorosis as a growing health hazard in Rajasthan and its control. *Acta Ecol.* 11;2: 59-67.

Schepers, J.S., Frank, K.D., and Watts, D.G. 1984. *Influence of irrigation and nitrogen fertilization on groundwater quality,* Proc. Int.Union of Geology and Geophysics, Hamburg, West Germany. 21-32.

Schindler, D.W. 1971. A hypothesis to explain the differences and similarities among lakes in the experimental lake area Noth Western Ontario. J.Fish. Res.Board, Can. 28: 295.

Schindler, D.W., and Fee, E.J. 1974. Experimental lake area: Whole lake experiments in eutrophication. *J. Fish.-Res. Board Can.* 31: 937-953.

Schindler, D.W., Kling, H., Schmidt, R.V., Prokopowich, J., Frost, V.E., Reid,R.A., and Capel, M. 1973. Eutrophication of lake 227 by addition of phosphate and nitrate: the second, third and fourth years of enrichment. 1978. 1971, 1972. J.Fish.Res.Board Can. 30: 1415-1440.

Schmidt, P., and Knetek, Z. 1970. *Epidemiological evaluation of nitrates as groundwater contaminants in Czechoslovakia.* Sixth International Water pollution conference, San-Francisco.

Seenayya, G. 1980. In *Progress in ecology.* Vol iv. eds, V.P. Aharwak and V.K.Sharma. *Today and Tomorrow,* New Delhi. 37-52.

Sengul, E., and Turkman, A. 1988. Kinetics of biochemical oxygen demand in waste stabilization ponds. In *Plants and pollutants in developed and developing countries.* Ed.M.A. Ozturk. Ege University, Turkey. 133-149.

Setchell, W.A. 1924. Ruppia and its environmental factors, *Botany,* 10: 286-288.

Shannon, E.E., and Brezonic, P.C. 1972. Eutrophication analysis: A multivariate approach. J. Sanitary Engg. Div. ASCE. 98 (S. A. Proc. Paper 8735), 37.

Sharma, K.C. 1986. Sources of water pollutants and physico-chemical characteristics of Anasagar, Ajmer. Paper presented at All India Seminar on Pollution in Western Rajasthan, Pali.

Sharma, P. 1986. Environmental upgradation of water bodies in urban areas of Madhya Pradesh. *Proc,Environment Day Conference,* EPCO, Bhopal. 36-59.

Sharma, R. 1977. Environmental problems - A bird eye view. In *Current trends* in *Indian environment.* Eds. Desh Bandhu and E.Chauhan. Today and Tomorrow, New Delhi. 25-35.

Shekhawat, S.S. 1983. Ecological study of Swaroopsagar lake with special reference to planktonic population and physico-chemical properties of water. Ph.D. Thesis, Sukhadia University, Udaipur.

Shih, C.S. 1976. A reliability assessment for regional water quality management. *Computers and Operational Research.* 3: 145-155.

Shrivastava, G.K., and Singh, V.P. 1990. Effect of municipal wastewater on yield and heavy metal contents of *Abelmoschus esculentus* (L) Moench, *Oikoassay,* 7: 5-7.

Shrivastava, V.M.S., Tripathi, R.S. , Srivastava, D.K., and Kumar, R. 1987. Accumulation and depletion of Chromium in Mystus vittatus (B1). *Acta Hydrochem.Hydrobiol.* 151 179-183.

Shrotriya, L. 1984. Water pollution studies in Ratlam area. M.Phil Thesis, Vikram University, Ujjain.

Shrotriya, L., and Dubey, P.S. 1989. Ecological assessment of river kurel (Ratlam). In *River pollution in India*. Ed. R.K.trivedi. Enviro-Scientific Publications, Karad.l-10.

Shrotriya, L., and Dubey, P.S. 1989. The management of aquatic ecosystems in some parts of western MP. In *Management of aquatic ecosystems*. The Academy of environmental biology, Muzaffarnagar. 281-285.

Shubinski, R.P., and Tierney, G.F. 1973. Effect of urbanization on water quality. *Proc. ASCE Urban Transportation division Conference on environmental* impact.p.180.

Shupe, J.L., and Alther, E.W. 1966. The effects of fluoride on livestock with particular reference to cattle. In *Handbook of environmental pharmacology*. Vol.20. Part I. Eds. O.Ethcler et al. Springer-Verlag, New York. 307-354.

Shuval, H.A. 1986. *Wastewater irrigation in developing countries: Health effects and technical solutions*. The World Bank. Washington DC.

Shuval, H.D., and Gruener, N. 1977. Infant methaemoglobinaemia and other health effects of nitrates in drinking water. In *Progress in water technology*. Pergamon Press, 8:183-194.

Siddiqui, A.H. 1972. Dental fluorosis in an Indian area with high natural fluoride content in water. Fluoride, 5:21-24.

Singh, D.K., Kumar, D.K., and Singh, V.P. 1985. Studies on pollutional effects of sugar mill and distillery effluents on seed germination and seedling growth of three varieties rice. *J. Environ. Biol.* 6:31-35.

Siva, K.M., and Ramamurthy, M.V. 1977. Indian J.Environ.Hlth. 19: 3: 199-209.

Somashekar, R.K., Gonda, M.T.G., Shettigar, S.L.N., and Srinath, K.P. 1984. Effect of industrial effluent on crop plants. *Indian J.Environ.Hlth.* 26:2:136-146.

Sreenivasan, A. 1963. Primary production in three upland lakes of Madras state, *India. Curr. Sci.* 32: 130-131.

Sreenivasan, A., and Sunderraj, R. 1967. *Indian J.Environ.Hlth.* 9: 13-21.

Sree Ramulu, U.S. 1990. Studies on the utilization of urban sewage and sludge for increasing crop production. *Proc. Indian Sci.Cong.Indore.*

Stangenberg, M. 1977. Some new views in the problem of hot water discharge to inland waters. In *Biological balance and thermal modifications.* Ed. M.Morois, Oxford, New York. 118-186.

Stevenson, F.J. 1986. *The internal cycle of nitrogen of soil: carbon, nitrogen, phosphorus, sulphur, micronutrients.* John Wiley & Sons, New York.

Storm, K.M. 1930. Limnological observations on Norwegian lakes. *Arch. Hydrobiol.* 21: 97-124.

Strauss, M. 1990. *The World* Hlth. *Forum.11*; 46-59.

Stumm, W., and Morgam, J.J. 1970. *Aquatic chemistry.* Wiley Interscience, New York.

Sundereshan, B.B., and Subramanyam, P.V.R. 1980. Industrial effluents - a threat to water resources. Paper presented at the seminar and paper meeting. Institute of Engineer, India.p.22.

Susheela, A.K. 1993. *Prevention and control* of *fluorosis in* India. Rajiv Gandhi National Drinking water Mission, Ministry of Rural Development, New Delhi. 20-22.

Sylvester, R.O. 1961. Nutrient content of drainage water from forested, urban, agricultural areas. Tech.Rep.Tuft.Sanit. Engg. *Centre. W-61-3*: 80-88.

Tailing, J.F., Wood, R. B., Prosser, M.V., and Baxter, R.M. 1973. The upper limit of photosynthetic productivity by plankton: evidence from Ethiopian soda lakes. *Freshwater Biol.* 3: 53-76.

Tam, N.F.Y., and Wong, Y.S. 1989. Wastewater nutrient removal by *Chloralla pyronoidosa and Scenedesmus* sp. Env. Poll. 58: 19-34.

Teal, J.M. 1977. Food chain transfer of hydrocarbons. In *Fate and effects of petroleum hydrocarbons in marine organisms and ecosystems.* Ed. D.A.Wolfe. Pergmon Press, New York.

Tehbutt, T.H.Y. 1973. *Water science and technology.* Barnes and Nobles, New York.

Tebbutt, T.H.Y. 1977. *Principles of water quality control.* Pergamon Press, Oxford.

Teotia, S.P.S., and Teotia, M. 1984. Endemic fluorosis in India; A challenging national health. *J.Assoc.Physic.India,* 32:4:347-352.

Thakur, U.C., and Deshpande, W.H. 1976. *Indian J.Environ.Hlth.* 18:2:136-148.

Thergaonkar, V.P., and Bhargava, R.K. 1974. Water quality and incidence of fluorosis in Jhunjhunu of Rajasthans: Preliminary observations. *Indian J. Environ, Hlth.* 16:2; 168-180.

Thienemann, A. 1918. Untersuchungen uber die Beziehungen zwischan dent Sauerst offgehalt der Wassers and der Zusammensetzung der Fauna in norddeutchen seen. *Arch Hydrobiol.* 12:1-16.

Thomas, E.A. 1969. The process of eutrophication in central European lakes. In *Eutrophication: causes, consequences and correctives.* Nat. Acad. Sci.Washington DC. 0.29.

Trivedi, R.C. 1980. Pollution studies of Chambal river and surroundings due to Nagda Industrial Complex. Ph.D. Thesis, Vikram University, Ujjain.

Trivedi, R.C. 1981. *Cent. Bd. Prev. Contr. Water Pollut.* Osmania University, Hyderabad. 175-188.

Trivedi, R.K. 1982. Souv.Nat.Conf. Environ. Kolhapur.E.A.K. 59-66.

Trivedi, R.K., and Joag, G.A. 1989. Treatment of industrial wastewater by water-hyacinth application. In *Pollution management in industries.* Ed. R.K.Trivedi. Env. Publications, Karad. 295-315.

Trivedi, R.K., Goel, P.K., Kulkarni, D.S., and Dharmadhikari, J.M. 1988. Studies on soil salinity in grape plantations areas in Tasgaon, Maharashtra. *Res. J. Pt. Environ.* 4:1:19-24.

Tsuchiya, K. 1969. *Kogai Special Issue,* 3-33.

Tyagi, O.D., and Mehra, M. 1990. *A text book of environmental chemistry.* Anmol Publications, New Delhi.

Tyagi, P.C., Basu, D.D., Chakrabarti, S.P., and Trivedi, R.C. 1991. Freshwater pollution and its control: The Indian experience. In *Aquatic sciences* In India. Eds. B.Gopal and V.Asthana. Indian Soc. for Limnology and Oceanography. New Delhi.

Ui, J. 1971. Rev. Interm. Oceanogr. Med. Tomes. 22 & 23: 6-30.

Ui, J. 1977. The world distribution of mercury contamination and their source of classification. *Kogai,* 1: 17-23.

Ui, J. 1980. Sewage system as appropriate technology. *Kogai,* 8: 1-24.

Uma, L., and Subramanian, G. 1990. Effective use of Cyanobacteria in effluent treatment. *Natl.Syrup.Cyanobacterial nitrogen fixation.* IARI New Delhi. 437-444.

Underwood, E.J. 1977. *Trace elements in human and animal nutrition.* Academic press. New York.

USEPA. 1972. *Water quality criteria.* US Environmental Protection Agency. R 3-73-033. Washington DC.

Uttormark, P.D., and Wall, P.J. 1975. *Lake classification - A trophic characterization of Wisconsin lakes*. Environmental Protection Agency (EPA - 660/3-75-033) Cornvallis, Oregon.

Vallentyne, J.R. 1973. The algal bowl - A faustian view of eutrophication. Proc.Fed.Ann.Soc.Exp.Biel.32:7:1754.

Varshney, C.K. 1981. Macrophytes as indicators of water quality. In *Proc.Workshop Biological indicators and indices* of *environmental pollution*. Ed. A.R.Zafar et al. Osmania University, Hyderabad. 53-59.

Vegis, A. 1948 (a). Einfluss tagesperiodisch wechselnder Aufbewahrungstemperatur *Stratiotes aloides*. Physiol.Plant. 1:216.235.

Vegis, A. 1948 (b). Einfluss der Aubewahrungstemperatur auf die Aktivitat der Knospen nach beendeter Wonterruhe. Physiol. Plant. 2: 117-130.

Vinberg, G.G. 1955. Significance of photosynthesis for oxygen enrichment of water during self purification of polluted waters. Trud. Vesesoyuz.Gidrobiol.Obshch.6146-69 (in Russian).

Viswanathan, R. 1957. *Indian J.Med. Res.* 45: Suppl. No.l.

Vora, B.B. 1975. *The soil and water management in India.*

Vyas, L.N. 1986. Ecology and conservation of lakes in and around Udaipur (Rajasthan). Final report MAB Prof. No. 20/46/81 - MAE/ENV-2. Bot. Dept. Sukhadia University, Udaipur.

Vyas, L.N., Sankhla, S.K., and Billore, D.K. 1981. Eutrophication in Udaipur lakes, evidence and prognosis. In *Integrated development of small and medium* towns; *Problems and strategic policy issues*. Eds. R.K.Wishwakarma and G.Jha. IIPA, New Delhi. 1-11.

Vyas, L.N., Sankhala, S.K., and Billore, D,K. 1982. Impact of human interference on accelerating pollution in Udaipur lakes. In *Advances in environmental research*. Ed.s.K. Agarwal, IEO, Kota. 45-52.

Vyas, L.N. 1989. State of the art of Pichhola - Fatehsagar wetland complex. In *Wetland, Mangroves and Biosphere reserves*. Ministry of environment and forests, New Delhi. 66-81.

Ward, R.C. 1973. *Principles of hydrology*. 2nd Edn. McGraw Hill, London.

Weber, C.A. 1907. Aufbau und vegetation der Moore Norddeut schlands. Beidl.Bot.Johrb. 90: 19-54.

Weber, J.A., and Nooden, L.D. 1976. Environmental and hormonal control of turion in *Myriophyllum verticillatum*. *Plant and Cell Physiol.* 17: 721-731.

Weibel, S.R., Anderson, R.J., and Woodward, R.L. 1964. J.Water Poll. *Contr.* 36s 7s 914.

Wetzel, R.G. 1968. Dissolved organic matter and phytoplankton productivity in marl lakes. Mitt. *Internat.* Verein.Limnol. 14: 261-270.

Wetzel, R.G. 1975. *Limnology.* W.B.Saunders Co., Philadelphia.

WHO, 1969. *Expert Committee on Amobiosis report.* World Health Organisation Technical Report Series No. 421.

WHO, 1972. *Health hazards of the human environment.* World Health Organisation, Geneva.

WHO, 1978. *Who Chronicle,* 32: 307-310. WHO. 1980. Tech.Rep.Ser.No.637 and 647.

WHO, 1984. *Guidelines for drinking water* quality.World Health Organisation, Geneva.

WHO, 1985. H*ealth hazards from nitrates in drinking water; Report on a WHO meeting.* Copenhagen.

WHO, 1987. *Guidelines for drinking water.* WHO, geneva.

Xintaras, C., Brug, J.R., Johnson, B.L., Tanaka, S., Lee, S.T., and Bender, J. 1979. Newrotoxic effects in workers exposed to Leptophos (phosvel). *Arh.hig.rada toksikol.* 30 Suppl. 553-592.

Yao, J.M. 1970. Regional water quality management. *J.Water Pollution Control Federation.* 3: 45: 407-411.

Zafar, A.R. 1959. Taxonomy of lakes. *Hydrobiol.* 13:287-299.

Zafar, A.R. 1964. Hydrobiol.23; 2: 179-195.

Zutshi, D.P., and Vass, K.K. 1977. *Trop. Ecol.*18:2:103-108.

INDEX